MOS Field-Effect Transistors and Integrated Circuits

MOS Field-Effect Transistors and Integrated Circuits

PAUL RICHMAN

Vice-President, Research and Development
Standard Microsystems Corporation

A WILEY-INTERSCIENCE PUBLICATION

JOHN WILEY & SONS, New York · London · Sydney · Toronto

Library of Congress Cataloging in Publication Data:

Richman, Paul.
    MOS field-effect transistors and integrated circuits.

    "A Wiley-Interscience publication."
    Includes bibliographies.
    1. Metal oxide semiconductors. 2. Field-effect transistors. 3. Integrated circuits. I. Title.

TK7871.85.R466   621.381'71   73-9892
ISBN 0-471-72030-5

Printed in the United States of America

10 9 8 7 6 5 4 3 2 1

To Ellen, Lee, and Alyson

Preface

Those of us who have been actively involved in the development and characterization of metal-oxide-silicon (MOS) field-effect devices since the early 1960s have felt a growing pride and pleasure as MOS technology has evolved from its early infancy to one of the most exciting and rapidly expanding fields in electronics today. During the course of the past 10 years, our focus has shifted from the first relatively unstable MOS capacitors and transistors to extremely complex large-scale-integrated (LSI) circuits, which in some instances contain more than 10,000 individual active transistors that have been selectively interconnected on a single monolithic silicon chip. Thus it has now become possible through the use of MOS technology to fabricate large self-contained electronic systems in a silicon area on the order of 0.04 in.^{2}. Recently, new processing techniques and new structural configurations have enabled MOSFETs to achieve higher operating speeds, much lower power dissipation, and greatly increased packing density. As a result, MOSFETs and integrated circuits are now having a pronounced effect on such diverse markets as random-access and read-only memories, portable electronic calculators, automotive electronics, data communications equipment, electronic wristwatches and clocks, and custom random-logic systems.

This book acquaints the reader with the basic semiconductor device physics and theory, predicted and experimentally observed electrical characteristics, and methods of fabrication of MOSFETs and provides an insight as to how specific changes in device parameters and the application of new processing techniques can substantially improve electrical performance in a number of different areas. The book, a graduate engineering text, will be of particular interest to students of both electrical engineering and semiconductor device physics. It also will be extremely useful to practicing engineers in the semiconductor industry as it provides in both theoretical and graphical form much of the essential information for the design and fabrication of MOSFETs and integrated circuits.

Chapter 1 introduces the reader to the field of MOSFETs. The basic surface field–effect is described, and the simple operation of the MOS capacitor and transistor structures is discussed. Chapter 2 treats the one-dimensional theory associated with the field-effect near the surface of a semiconductor and shows how this theory can be extended to predict the behavior of the threshold voltage of a MOS device as a function of oxide thickness, substrate doping concentration, interface charge density, and metal-semiconductor work function difference. Finally, the effect of an applied substrate voltage on the threshold voltage of a MOSFET is discussed. Chapter 3 is devoted entirely to the theory of the MOS capacitor structure. The capacitance observed between gate and substrate is shown to be a strong function of device parameters, applied gate voltage, and frequency of operation. The theoretical behavior of the MOS capacitor under inversion, depletion, deep-depletion, and accumulation conditions is treated for both high- and low-frequency gate signals, and the results are compared with typical experimental data. Chapter 4 gives a detailed description of the three-terminal characteristics of MOSFETs as a function of the applied gate and drain voltages. Operation below pinch-off, in the region of saturated drain current, and in the breakdown region are each treated separately, and the physical mechanisms responsible for the observed electrical characteristics in each region are related to the device theory. Chapter 5 deals with the effect of temperature variations on the electrical characteristics of MOSFETs. Chapter 6 examines the properties of the silicon-silicon dioxide system and, in particular, discusses the characteristics of the silicon-silicon dioxide interface and how these characteristics affect the properties of silicon MOS devices. Finally, Chapter 7 treats the speed limitations associated with MOSFETs and describes how MOS structures can be fabricated to achieve extremely high-speed operation. An equivalent circuit for a MOSFET is developed that predicts the high-frequency behavior of the device. The maximum frequency of operation is related to transit-time considerations within the source-drain gap, and it is shown that the maximum frequency of operation of any MOSFET is proportional to its transconductance divided by the input capacitance looking into the gate electrode. The rest of the chapter examines the different techniques that have been developed over recent years to achieve high-speed MOS devices. In addition to the cited references, each chapter also contains a supplemental list of references and a number of problems to test one's understanding of the material treated.

PAUL RICHMAN

Hauppauge, New York
August 1973

Contents

MOS Field-Effect Transistors and Integrated Circuits

1

Introduction

If a highly conductive electrode is brought in very close proximity to the surface of a lightly doped semiconductor substrate, a voltage applied to the electrode with respect to the substrate can have a pronounced effect on the electronic properties of the semiconductor surface. In particular, if n-type silicon is used as the substrate material, the application of a large positive voltage to the conducting electrode will attract an increased concentration of negative mobile charges (electrons) to the silicon surface and the surface will become *accumulated*. On the other hand, if a negative voltage is applied to the electrode, the electrons will be repelled from the silicon surface, and the region near the surface will become totally *depleted* of mobile charge, although nonmobile positive charge will remain there in the form of the ionized donor-type impurities. If the magnitude of the applied negative voltage is increased, mobile positive charges (holes), which are minority carriers within the n-type silicon, will be attracted to the surface, forming a conducting layer in which the holes will be majority carriers. In this case, the silicon surface is said to be *inverted*. Similarly, if the silicon substrate is p-type, a surface accumulation layer will form with the application of a large negative voltage to the electrode, while a surface inversion layer will form when the applied voltage is positive. The simple phenomenon described above is commonly referred to as the *surface field effect* and is the underlying principle behind the operation of metal-oxide-semiconductor (MOS) field-effect transistors (FETs) and other related insulated-gate electronic devices.

There are two distinctly different types of MOSFETs. n-channel MOSFETs operate by means of surface conduction of electrons under

1

the influence of a transverse applied voltage, while the charge carriers that give rise to current flow in a p-channel device are holes. Because the mobility of electrons in silicon is appreciably greater than that of holes for equal substrate doping concentrations, n-channel MOS transistors are generally faster than p-channel devices that have been fabricated in a similar manner. The cross-sectional structure of a p-channel MOSFET is shown in Figure 1.1. The device consists of two closely spaced, degenerately doped p^+ regions, the "drain" and the "source," which have been diffused into a lightly doped n-type silicon substrate. Typically, the distance separating the two diffused regions is on the order of a few tenths of a mil. A thin insulating layer of silicon dioxide is formed directly above the region separating the two p^+ diffusions by oxidizing the surface of the silicon at high temperatures in an oxygen-rich ambient. Metal contacts, usually aluminum, are made to both diffused regions, and a "gate" electrode is positioned directly over and completely covering the region between the drain and source. As previously mentioned, the gate electrode must be of a highly conductive material. Usually, aluminum is also used for the gate, but a great deal of work has been done using both highly doped polycrystalline silicon and other metals such as gold, titanium, platinum, nickel, and chromium. Although the gate electrode need not be made of metal and the insulator need not be an oxide, the term metal-oxide-semiconductor (MOS) generally relates to the more conventional gate structure which consists of a metal electrode which is separated by a thin layer of oxide from the underlying semiconductor substrate. The structure of an n-channel MOSFET is similar to the device shown in Figure 1.1 except that n^+ regions are diffused into a p-type silicon substrate.

There is no physical distinction between the drain and source regions in a MOSFET because of the inherent symmetry of the structure. Rather, the biasing conditions generally determine which region is considered the source and which region is considered the drain. For a p-channel MOS transistor,

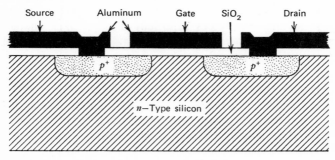

FIGURE 1.1   Cross-sectional view of a p-channel mosfet.

the p^+ region with the most positive potential is defined as the source. Similarly, the source of an n-channel device is defined as the n^+ region with the most negative potential. In most applications, the substrate and source of a MOSFET are usually kept at the same potential.

Most p-channel MOSFETs are usually *enhancement-type* devices. That is, there will be no current flow between the drain and source when a negative voltage is applied to the drain, with respect to the source, and the applied gate-to-source voltage is set equal to zero. Referring once again to Figure 1.1, no conducting channel will be present at the silicon surface in the region between the two p^+ diffusions with zero gate voltage, and when a negative drain-to-source voltage is applied, no transverse current will flow through the structure because the drain junction will be reverse biased. However, if a large negative voltage is applied to the gate electrode with respect to the source, a p-type surface inversion layer will form in the silicon directly below the gate, providing a conducting *channel* between the drain and the source which results in appreciable current flow between the two regions. Consequently, it can be seen that a p-channel enhancement-type MOSFET will be "normally off" when the gate voltage is equal to zero but can be turned "on" with the application of a negative gate voltage. Such a device is extremely applicable to digital switching applications.

The gate-to-source voltage which is required to achieve surface inversion and, as a result, conduction between the drain and source regions, is defined as the *threshold voltage* of the transistor. In normally off enhancement-type MOSFETs, the threshold voltage is a negative quantity for p-channel configurations and a positive quantity for n-channel configurations. However, n-channel MOSFETS fabricated on lightly doped p-type silicon substrates are usually "normally on" with zero gate voltage. Such devices are called *depletion-type* because their conductance can be "depleted" by the application of a gate voltage of opposite polarity to that of the drain voltage. Thus only the application of a negative gate-to-source voltage will turn an n-channel depletion-type MOSFET "off"; hence the threshold voltage of the device will be negative. The n-channel MOS transistors are usually of the depletion type because of the existence of a positive layer of fixed charge which is located in the silicon dioxide near the silicon surface. This fixed positive charge layer has been found to exist in devices fabricated on both n-type and p-type silicon substrates. Since the positive charge density in the oxide attracts an equal and opposite amount of charge in the silicon directly below, an accumulation layer tends to form at the surface of an n-type silicon substrate, while an n-type surface inversion layer tends to form on a lightly doped p-type substrate. (Although a surface inversion layer will not form if the doping concentration in the p-type substrate is sufficiently high, the effect of the positive oxide charge will be to reduce the concentration of mobile positive charge carriers (holes) near the silicon surface.) It is the existence of a surface inversion layer which allows current to flow between drain and source with zero gate voltage in a depletion-type device.

It can be readily seen that the MOSFET, because of its insulated gate electrode, is a voltage-controlled device rather than a current-amplifying structure like the conventional bipolar junction transistor, which relies on a small base-to-emitter current to control much larger amounts of collector-to-emitter current flow. Because of the extremely high input impedance associated with the gate electrode of the MOS transistor, it has been frequently referred to as a solid-state analog of the vacuum-tube triode. The inherent difference between the two devices is that the gate electrode of the MOS transistor actually modulates the conductivity of the semiconducting region between the two current-carrying electrodes, the drain and source, while the grid of the vacuum-tube triode sets up a retarding potential field that impedes the flow of electrons traveling between the cathode and the anode. The three-terminal electrical characteristics of the MOSFET are also typically quite different from those of the vacuum-tube triode. When the drain current is plotted versus the applied drain-to-source voltage for varying values of gate-to-source voltage, the observed characteristics usually exhibit current saturation at values of drain voltage approximately equal to the gate voltage minus the threshold voltage; the resulting device characteristics are typically pentode-like.

In general, the three-terminal electrical characteristics of MOSFETs can be divided into three regions. At values of applied drain voltage which are sufficiently small as to be very much less than the magnitude of the gate voltage minus the threshold voltage, the drain current at constant gate voltage is found to increase linearly with increasing drain voltage, and the characteristics of the device are like those of a voltage-variable resistor, with the drain-to-source resistance decreasing steadily with increasing values of applied gate-to-source potential. Under these conditions, the MOSFET is said to be operating in the *variable-resistance region*. As the applied drain-to-source voltage is increased and becomes larger than the gate voltage minus the threshold voltage, the drain current tends to saturate and becomes relatively constant and independent of the drain voltage. The saturation of the drain current is a direct result of the formation of a depletion region near the drain end of the conducting surface channel. When the device is operating under these conditions, it is said to be operating in the *region of saturated current flow*. An interesting characteristic of this region is that the drain current after saturation is approximately proportional to the square of the gate voltage minus the threshold voltage. At very large values of applied drain voltage, avalanche breakdown of the drain diode occurs and the drain current begins to increase extremely rapidly with increasing drain voltage.

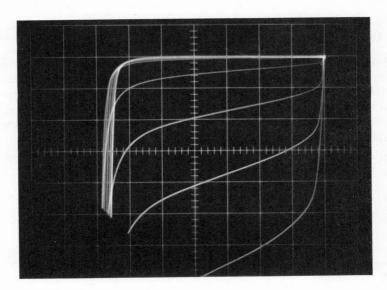

FIGURE 1.2 Three-terminal current-voltage characteristics associated with a typical *p*-channel enhancement-type MOSFET; vertical scale: drain current: -1 mA/div.; horizontal scale: drain voltage: -10 V/div.; gate voltage: 0 to -8 V, in -1 V steps; substrate voltage: 0 V.

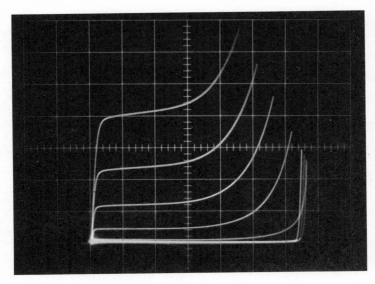

FIGURE 1.3 Three-terminal current-voltage characteristics associated with a typical *n*-channel enhancement type MOSFET; vertical scale: drain current: $+500\ \mu$A/div.; horizontal scale: drain voltage: $+10$ V/div.; gate voltage: 0 to $+7$ V, in $+1$ V steps; substrate voltage: 0 V.

At this time, the device is said to be operating in the *avalanche breakdown region*. All three regions of MOSFET operation are treated in detail in later chapters. Typical current-voltage characteristics for p- and n-channel enhancement-type devices are shown in Figures 1.2 and 1.3. For each MOSFET, the three individual regions of operation described above are clearly evident.

2

The Field Effect at the Surface of a Semiconductor

2.1 THE SPACE CHARGE REGION, MOBILE CARRIER CONCENTRATION, AND ELECTRIC FIELD INTENSITY NEAR THE SURFACE OF A SEMICONDUCTOR

Through the use of a one-dimensional solution to Poisson's equation, the relationships between the space charge region, mobile carrier concentration, and electric field intensity near the surface of a semiconductor that is under the influence of an applied perpendicular electric field can be calculated.[1] Referring to the simple MOS capacitor structure shown in Figure 2.1, if $x = 0$ is defined to be at the silicon-silicon dioxide interface, and if it is also assumed that all variations in the applied gate voltage are slow enough so that the entire system always remains in *thermal equilibrium*, then the electrostatic potential in the silicon, ϕ, can be related to the total space charge density per unit volume, ρ, by Poisson's equation:

$$\frac{\partial^2 \phi}{\partial x^2} = - \frac{\rho}{\epsilon_s},\qquad(2.1)$$

where ϵ_s is the dielectric constant of the silicon substrate. The electric field intensity $\mathscr{E}_x$ is related to the electrostatic potential by

$$\mathscr{E}_x = - \frac{\partial \phi}{\partial x}.\qquad(2.2)$$

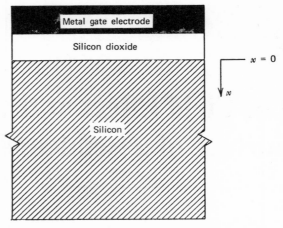

FIGURE 2.1   Simple MOS capacitor structure.

The total space charge density in the semiconductor as a function of x can be written as the sum of the ionized donor concentration per unit volume, N_D^+, and the hole concentration per unit volume, p, minus the sum of the ionized acceptor and electron concentrations per unit volume, N_A^- and n, respectively, all times the electronic charge, q. That is,

$$\rho = q(N_D^+ - N_A^- + p - n). \qquad (2.3)$$

Deep in the bulk of the silicon substrate, both the electrostatic potential and the electric field are approximately equal to zero and a condition of charge neutrality must exist. Thus, for very large values of x,

$$N_D^+ - N_A^- = n - p. \qquad (2.4)$$

If both the Fermi energy and the Fermi level are measured in the conventional manner, from the center of the forbidden band gap, then for very large values of x deep in the silicon bulk, the free carrier densities will be given by

$$n = n_i \exp\left(\frac{-q\phi_F}{kT}\right),$$

and $\qquad (2.5)$

$$p = n_i \exp\left(\frac{+q\phi_F}{kT}\right),$$

where ϕ_F is the Fermi potential, n_i the intrinsic carrier concentration of silicon, k Boltzmann's constant, and T the temperature in degrees Kelvin.

As previously mentioned, the electrostatic potential for large values of x will be approximately equal to zero. The value of electrostatic potential at

$x = 0$ will be denoted by ϕ_s. The quantity $\phi_s = \phi$ at $x = 0$ is commonly referred to as the *surface potential*. If, for example, the case of a p-type silicon substrate is considered, as shown in Figure 2.2, ϕ_s will increase with increasingly positive gate voltage and can be seen to be a direct measure of the amount of band-bending occurring at the surface of the silicon as a function of the applied gate potential.

If all the impurities in the substrate are assumed to be ionized, which is true for silicon at room temperature and above, then combining (2.4) and (2.5), with $N_D^+ = N_D$ and $N_A^- = N_A$, yields

$$N_D - N_A = 2n_i \sinh \left(\frac{-q\phi_F}{kT} \right). \tag{2.6}$$

Near the surface of the silicon, where the electrostatic potential ϕ is no longer equal to zero, the electron and hole concentrations per unit volume are respectively given by

$$n = n_i \exp \left(\frac{-q(\phi_F - \phi)}{kT} \right)$$

and $\tag{2.7}$

$$p = n_i \exp \left(\frac{+q(\phi_F - \phi)}{kT} \right).$$

Thus it follows that for any value of electrostatic potential in the silicon, the difference between the electron and hole concentrations per unit volume

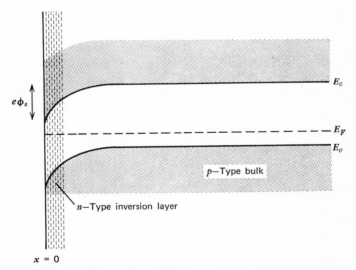

FIGURE 2.2   Band-bending at the surface of a p-type semiconductor.

will be given by

$$n - p = 2n_i \sinh\left(\frac{q(\phi - \phi_F)}{kT}\right). \tag{2.8}$$

Through the use of (2.1), (2.3), (2.6), and (2.8), Poisson's equation can now be rewritten as

$$\frac{\partial^2 \phi}{\partial x^2} = -\frac{2qn_i}{\epsilon_s}\left\{\sinh\left(\frac{-q\phi_F}{kT}\right) - \sinh\left[\frac{q(\phi - \phi_F)}{kT}\right]\right\}. \tag{2.9}$$

If, for simplicity, the following quantities are defined:

$$u \equiv \frac{q\phi}{kT}$$

$$u_F \equiv \frac{q\phi_F}{kT} \tag{2.10}$$

$$u_s \equiv \frac{q\phi_s}{kT}$$

and since $\sinh(-u_F) = -\sinh(u_F)$, then (2.9) becomes

$$\frac{\partial^2 u}{\partial x^2} = +\frac{2q^2 n_i}{\epsilon_s kT}[\sinh(u - u_F) + \sinh(u_F)]. \tag{2.11}$$

The intrinsic Debye length is defined as

$$L_D \equiv \left(\frac{\epsilon_s kT}{2n_i q^2}\right)^{1/2}. \tag{2.12}$$

Graphs of the variation of the intrinsic carrier concentration and the intrinsic Debye length, for silicon, as a function of temperature are shown in Figures 2.3 and 2.4, respectively.

Using the definition of the intrinsic Debye length given above, Poisson's equation can be written as

$$\frac{\partial^2 u}{\partial x^2} = \frac{1}{L_D^2}[\sinh(u - u_F) + \sinh(u_F)]. \tag{2.13}$$

Integrating from deep in the bulk of the substrate (where $(\partial u/\partial x) = 0$) toward the surface of the silicon gives

$$\int_0^{\partial u/\partial x}\left(\frac{\partial u}{\partial x}\right)d\left(\frac{\partial u}{\partial x}\right) = \frac{1}{L_D^2}\int_0^u[\sinh(u - u_F) + \sinh(u_F)]\,du \tag{2.14}$$

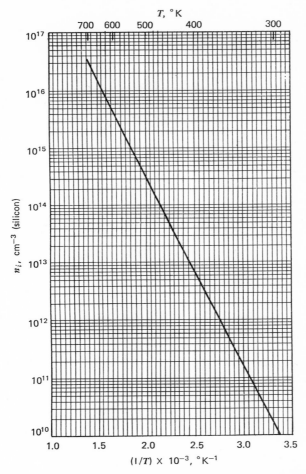

FIGURE 2.3   Behavior of the intrinsic carrier concentration n_i in silicon versus temperature in degrees Kelvin.

or

$$\frac{1}{2}\left(\frac{\partial u}{\partial x}\right)^2 = \frac{1}{L_D{}^2}\left[\cosh\left(u - u_F\right) - \cosh\left(u_F\right) + u\sinh\left(u_F\right)\right]. \quad (2.15)$$

From (2.2), the electric field in the silicon in the x direction is given by

$$\mathscr{E}_x = -\frac{\partial\phi}{\partial x} = -\frac{kT}{q}\left(\frac{\partial u}{\partial x}\right).$$

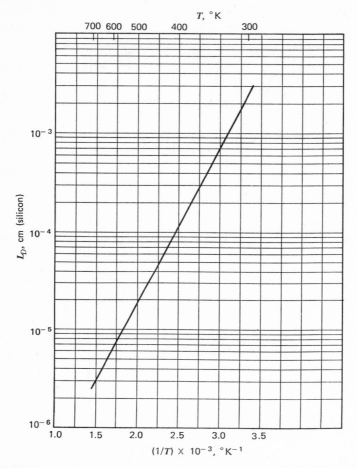

FIGURE 2.4 Behavior of the intrinsic Debye length L_D in silicon versus temperature in degrees Kelvin.

Therefore, taking the negative root yields

$$\mathscr{E}_x = + \frac{kT}{q} \left\{ \frac{2}{L_D{}^2} \left[\cosh (u - u_F) - \cosh (u_F) + u \sinh (u_F) \right] \right\}^{1/2}. \quad (2.16)$$

At the silicon surface $x = 0$, $u = u_s$ and the electric field is

$$\mathscr{E}_s = \frac{kT}{qL_D} \left\{ 2[\cosh (u_s - u_F) - \cosh (u_F) + u_s \sinh (u_F)] \right\}^{1/2}. \quad (2.17)$$

By Gauss' law, the *total* charge density per unit area in the silicon needed to terminate this field is

$$Q_{\text{total}} = -\epsilon_s \mathscr{E}_s,$$

where

$$Q_{\text{total}} \equiv Q_{\text{inv}} + Q_{SD}.$$

$$(2.18)$$

Q_{inv} is the charge density per unit area contained in the surface inversion layer, and Q_{SD} is the charge density per unit area contained in the surface depletion layer. Use of (2.12), (2.17), and (2.18) yields an expression for the total surface space charge per unit area for a *p*-type silicon substrate:

$$Q_{\text{total}} = -2qn_i L_D \{2[\cosh(u_s - u_F) - \cosh(u_F) + u_s \sinh(u_F)]\}^{1/2} \quad (2.19)$$

The surface space charge density is negative for the *p*-type substrate and positive for an *n*-type substrate.

If the applied gate-to-substrate voltage is adjusted so that an inversion layer has not yet formed at the surface of the silicon and the surface is actually intrinsic (i.e., $\phi_s = \phi_F$), then $Q_{\text{inv}} = 0$ and the charge density in the surface depletion region becomes

$$Q_{SD}|_{\phi_s = \phi_F} = -2qn_i L_D \{2[1 - \cosh(u_F) + u_F \sinh(u_F)]\}^{1/2}. \quad (2.20)$$

For gate voltages sufficiently large to result in surface inversion, the charge density in the (*n*-type) inversion layer can be written as:[2]

$$Q_{\text{inv}} = Q_n = -q \int_{x=0}^{x=x_i} n(x) \, dx = -q \int_{u=u_s}^{u=u_F} \frac{n(u) \, du}{(du/dx)}$$

$$= +q \int_{u=u_F}^{u=u_s} \frac{n(u) \, du}{(du/dx)}, \quad (2.21)$$

where x_i is defined as the point in the silicon substrate where $\phi_i = \phi_F$ and the silicon is intrinsic. Note that $u_s = u(x = 0)$ and $u_F = u(x = x_i)$. From (2.7) and (2.10), the quantity $n(u) \, du$ can be expressed as

$$n(u) \, du = n_i e^{(u - u_F)} \, du. \quad (2.22)$$

Substituting $\mathscr{E}_x = -kT/q(\partial u/\partial x)$ into (2.16) yields

$$\frac{du}{dx} = -\frac{1}{L_D} \{2[\cosh(u - u_F) - \cosh(u_F) + u \sinh(u_F)]\}^{1/2}. \quad (2.23)$$

Thus the charge density in the (*n*-type) inversion layer per unit area may be obtained by combining (2.21), (2.22), and (2.23). The result is

$$Q_n = -qn_i L_D \int_{u=u_F}^{u=u_s} \frac{e^{(u - u_F)} \, du}{\{2[\cosh(u - u_F) - \cosh(u_F) + u \sinh(u_F)]\}^{1/2}}. \quad (2.24)$$

A similar expression is obtained for Q_p, the (positive) charge density per unit area associated with a p-type surface inversion layer on an n-type substrate. Since the total charge density in the surface space charge region is given by the sum of the charge densities in the inversion layer and the depletion region,

$$Q_{SD} = Q_{total} - Q_n, \qquad (2.25)$$

where Q_{total} is given by (2.19) and Q_n is given by (2.24). Grove, Deal, Snow, and Sah have shown that, for high values of surface potential, practically all of the charge density associated with Q_{total} lies within the inversion layer.[2] That is,

$$\lim_{u_s \gg u_F} Q_n \to Q_{total}. \qquad (2.26)$$

Physically, this means that under equilibrium conditions, once an inversion layer is formed, any additional surface charge that results from increasing the applied gate voltage will lie within the inversion layer and, consequently, will increase the surface conductance. Hence the depletion charge density Q_{SD} will tend to saturate to a constant value after inversion takes place. It follows that the width of the surface depletion region will also approach a constant (maximum) value after the surface has become inverted.

If it is assumed that the charge density per unit area contained in the surface depletion region saturates at a maximum value, $Q_{SD_{max}}$ after *strong inversion* has taken place[3] with $\phi_s \cong 2\phi_F$, and if it is also assumed that all the charge in the surface space charge layer lies within the depletion region until this saturation takes place, then substitution in (2.19) yields

$$Q_{SD_{max}} \equiv Q_{total}\big|_{\phi_s \cong 2\phi_F} \cong -2qn_iL_D\left[\frac{4q\phi_F}{kT}\sinh\left(\frac{q\phi_F}{kT}\right)\right]^{1/2}. \qquad (2.27)$$

Under typical conditions such that $(q\phi_F/kT) > 1$ and $\exp(q\phi_F/kT) \ggg 1$, then

$$\sinh\left(\frac{q\phi_F}{kT}\right) \cong \tfrac{1}{2}\exp\left(\frac{q\phi_F}{kT}\right),$$

and use of (2.12) and (2.27) gives

$$Q_{SD_{max}} \cong -\left[\left(\frac{2kTqn_i\epsilon_s}{q}\right)^{1/2}\right]\left\{\left[\frac{2q\phi_F\exp(q\phi_F/kT)}{kT}\right]^{1/2}\right\}, \qquad (2.28)$$

and since $p = n_i\exp(q\phi_F/kT) \cong N_A$, (2.28) reduces to

$$Q_{SD_{max}} \cong -(4q\epsilon_sN_A\phi_F)^{1/2}. \qquad (2.29)$$

The maximum width of the surface depletion region after strong inversion takes place is related to $Q_{SD_{max}}$ by

$$Q_{SD_{max}} = -qN_A x_{d_{max}}. \tag{2.30}$$

Therefore, the maximum width of the surface depletion region is given by

$$x_{d_{max}} \cong \left(\frac{4\epsilon_s \phi_F}{qN_A}\right)^{1/2} = \left[\frac{2\epsilon_s \phi_s (\text{strong inversion})}{qN_A}\right]^{1/2}. \tag{2.31}$$

In summary, for a MOS structure, formed on a p-type silicon substrate, operating under equilibrium conditions, if the applied gate voltage is such that only a surface depletion region exists, then the depletion region charge density per unit area is given by

$$Q_{SD} = -2qn_i L_D \{2[\cosh(u_s - u_F) - \cosh(u_F) + u_s \sinh(u_F)]\}^{1/2}. \tag{2.32}$$

This can also be expressed as

$$Q_{SD} = -qN_A x_d, \tag{2.33}$$

where x_d is the width of the surface depletion region:

$$x_d = \frac{2n_i L_D}{N_A} \{2[\cosh(u_s - u_F) - \cosh(u_F) + u_s \sinh(u_F)]\}^{1/2}. \tag{2.34}$$

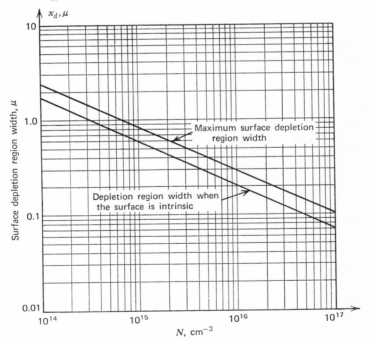

FIGURE 2.5 Variation of surface depletion region width as a function of the effective doping concentration in the silicon.

For gate voltages such that the surface potential is greater than or equal to $2\phi_F$, and the surface of the silicon is strongly inverted, $Q_{SD} \rightarrow Q_{SD_{max}}$ and $x_d \rightarrow x_{d_{max}}$. Then

$$Q_{total} = Q_{SD_{max}} + Q_n \qquad (2.35)$$

where $Q_{SD_{max}}$ is approximately given by (2.29) and Q_n is approximately given by (2.24). Following the very same procedure which is outlined above for a p-type substrate, similar expressions can be derived for Q_{SD}, x_d, Q_p, $Q_{SD_{max}}$, and $x_{d_{max}}$ for the case of a MOS structure fabricated on n-type material.

If the difference in effective mass between the conduction and valence bands is neglected, and the absolute value of the Fermi potential in the silicon substrate is approximated by

$$|\phi_F| = \frac{kT}{q} \ln \left(\frac{N}{n_i}\right), \qquad (2.36)$$

where N is the effective impurity concentration in the substrate per unit volume ($N = |N_D - N_A|$) then graphs of $x_{d_{max}}$ and $|Q_{SD_{max}}|$ can be plotted versus N for MOS structures with either p- or n-type silicon substrates. These graphs are shown in Figures 2.5 and 2.6.

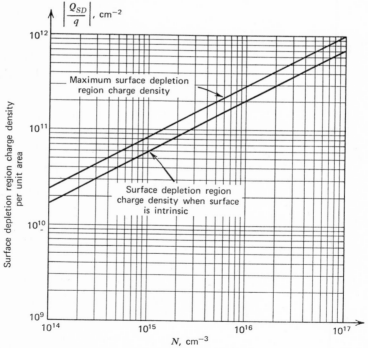

FIGURE 2.6 Variation of surface depletion region charge density per unit area as a function of the effective doping concentration in the silicon.

2.2 THE EFFECT OF FIXED POSITIVE OXIDE CHARGE, INSULATOR THICKNESS, SUBSTRATE RESISTIVITY, AND GATE ELECTRODE WORK FUNCTION ON THE THRESHOLD VOLTAGE OF A MOS TRANSISTOR

When a gate voltage is applied to a MOSFET that is connected with both its source and substrate at a common potential, the resultant electric field is confined, for all practical purposes, to a volume consisting of the gate electrode, the gate insulator, and the surface space charge region in the silicon. It follows, from Gauss' law, that a condition of charge neutrality must exist, under equilibrium conditions, in this volume. By employing the requirement for charge neutrality, theoretical threshold voltages for both p-channel and n-channel MOSFETs can be predicted. Also, graphical relationships can be obtained that indicate the dependence of MOS threshold voltages on the fixed positive charge density located in silicon dioxide layers in close proximity to the silicon surface, the thickness of the gate insulator, the resistivity of the silicon substrate, and the difference in work functions between the material used for the gate electrode and the silicon substrate itself.[4]

2.2.1 Threshold Voltages for n-channel MOS Devices

The charge distribution, under equilibrium conditions, for an n-channel MOSFET with a silicon dioxide gate insulator which has been fabricated on a uniformly doped p-type silicon substrate is shown in Figure 2.7. Figure 2.7a illustrates the situation at low gate voltages where only a depletion region, consisting of nonmobile ionized acceptors extending to $x = x_d$, is present at the surface. Figure 2.7b illustrates the situation in which the gate voltage is sufficiently high to invert the surface of the p-type substrate and to form an n-type inversion layer. As previously discussed, when this takes place, the depletion region has reached its maximum width, $x = x_{d_{max}}$. Then Q_G is defined as the charge density present on the gate electrode per unit area as a function of the applied gate voltage; Q_{SS} is the fixed positive surface state charge density per unit area located in the silicon dioxide in very close proximity to the oxide-silicon interface; Q_n is the charge density contained in the surface inversion layer per unit area and Q_{SD} and $Q_{SD_{max}}$ are the charge density, and maximum charge density, per unit area contained in the surface depletion region. The threshold voltage V_T will be defined as the gate voltage required to achieve strong inversion, with $\phi_s = 2\phi_F$.

Once again, making the assumption that all the charge in the surface space charge layer resides in the depletion region when the surface potential is equal to twice the Fermi potential, and that all *additional* negative charge attracted to the silicon surface through the application of increasingly positive

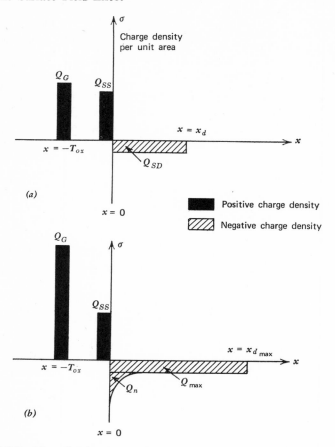

FIGURE 2.7 The charge distribution in an n-channel MOSFET (a) before inversion; (b) after inversion.

gate voltage will become part of the inversion layer, then it follows that when $\phi_s = 2\phi_F$, $Q_n = 0$, and the condition for charge neutrality is

$$Q_G + Q_{SS} + Q_{SD_{max}} = 0. \qquad (2.37)$$

In actual MOS structures, the voltage across the oxide layer, V_{OX}, will differ from the applied gate-to-substrate voltage V_G because of the difference between the metal-silicon dioxide and silicon dioxide-silicon barrier energies and because of the band-bending at the silicon surface.[2] The charge density Q_G present on the gate electrode per unit area can be determined by examining the energy-band diagram of a MOS structure, which is shown in Figure 2.8 for the case in which the applied gate voltage is set equal to zero. The

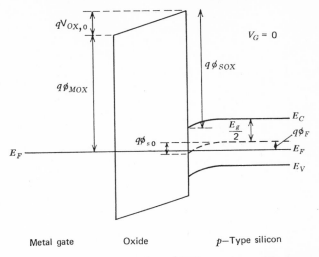

FIGURE 2.8 Energy-band diagram for a MOS structure with zero applied gate voltage.

metal-silicon dioxide barrier energy is denoted by ϕ_{MOX} and the energy associated with the silicon-silicon dioxide barrier is denoted by ϕ_{SOX}. The voltage across the oxide layer with zero gate voltage is denoted by $V_{OX,0}$ and the value of the surface potential with zero gate voltage is denoted by ϕ_{s0}. The silicon band-gap energy E_g is merely the difference between the energies associated with the conduction and valence bands in the silicon substrate. As shown in Figure 2.8, when the applied gate voltage is equal to zero, the position of the Fermi level is constant throughout the structure. Thus by summing the energies on both sides of the barrier presented by the silicon dioxide gate insulator, it follows that

$$q\phi_{MOX} + qV_{OX,0} = q\phi_{SOX} + \frac{E_g}{2} - q\phi_{s0} + q\phi_F. \qquad (2.38)$$

If the quantity $\phi_{MS'}$ is defined as the metal-semiconductor work function difference, such that

$$\phi_{MS'} \equiv \phi_{MOX} - \phi_{SOX} - \frac{E_g}{2q} - \phi_F, \qquad (2.39)$$

then, combining (2.38) and (2.39), the total voltage across the oxide layer with $V_G = 0$ is

$$V_{OX,0} = -(\phi_{MS'} + \phi_{s0}). \qquad (2.40)$$

It also follows that when a voltage is applied to the gate electrode, the voltage across the oxide can be related to the gate voltage by

$$V_G = V_{OX} - V_{OX,0} + \phi_s - \phi_{s0}. \qquad (2.41)$$

Thus in terms of the voltage across the oxide layer, the metal-semiconductor work function difference, and the surface potential, the gate voltage can be expressed as

$$V_G = V_{OX} + \phi_{MS'} + \phi_s. \qquad (2.42)$$

Deal, Snow, and Mead[5] measured the barrier heights associated with both the silicon-silicon dioxide interface and the interfaces between various metal electrodes and silicon dioxide. They found the silicon-silicon dioxide barrier energy ϕ_{SOX} to be approximately independent of the orientation or conductivity type of the silicon and equal to 4.35 eV. The results of their measurements of metal-silicon dioxide barrier heights are shown in Table 2.1. The data were obtained using both photoemission and MOS capacitance-voltage techniques.

$\phi_{MS'}$ can be computed as a function of the conductivity type and the effective impurity doping concentration of the silicon substrate through the use of (2.39) in conjunction with the measured values of ϕ_{MOX} and ϕ_{SOX} and the values of Fermi potential as a function of the substrate doping concentration obtained from either Figure 2.9 or 2.10, depending on the conductivity type of the substrate. Values of $\phi_{MS'}$ calculated in this manner for MOS structures with *aluminum* gate electrodes are shown in the graph of Figure 2.11.

Now the charge density on the gate electrode per unit area can be written as

$$Q_G = C_{ox}V_{OX} = \frac{\epsilon_{ox}V_{OX}}{T_{ox}}, \qquad (2.43)$$

where C_{ox} is the dielectric capacitance per unit area across the insulating gate oxide. At the threshold voltage, $V_G = V_T$ and, by definition, $\phi_s = 2\phi_F$.

TABLE 2.1   Metal-Silicon Dioxide Barrier Energies (After Deal, Snow, and Mead[5])

Metal	ϕ_{MOX} (photo)	$\phi_{MOX(C-V)}$
Aluminum	3.2 eV	3.2 eV
Copper	3.8 eV	3.75 eV
Gold	4.1 eV	4.05 eV
Magnesium	2.25 eV	2.5 eV
Nickel	3.7 eV	3.7 eV
Silver	4.15 eV	4.1 eV

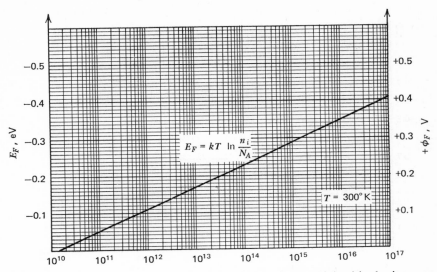

FIGURE 2.9 Variation of Fermi level and Fermi potential with doping concentration in p-type silicon at room temperature (measured from the center of the forbidden band gap).

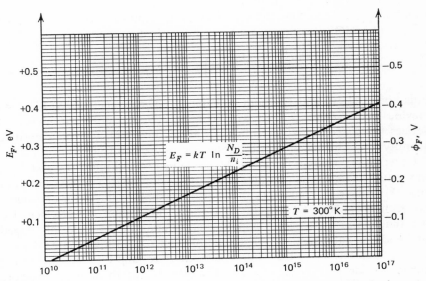

FIGURE 2.10 Variation of Fermi level and Fermi potential with doping concentration in n-type silicon at room temperature (measured from the center of the forbidden band gap).

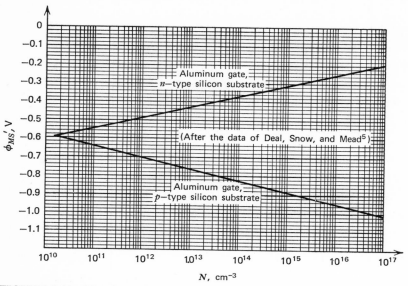

FIGURE 2.11   Metal-semiconductor work function difference as a function of the doping concentration in the silicon substrate, for an aluminum gate electrode.

Combining (2.37), (2.42), and (2.43) yields an expression for the threshold voltage of a MOSFET fabricated on a uniformly doped substrate:

$$V_T = \left(\frac{-Q_{SS} - Q_{SD\text{max}}}{\epsilon_{ox}}\right) T_{ox} + \phi_{MS'} + 2\phi_F. \tag{2.44}$$

Equation 2.44 is valid for both n-channel and p-channel MOSFETs. As previously discussed, $Q_{SD_{\text{max}}}$ will be a negative quantity for n-channel devices fabricated on p-type substrates and, conversely, will be a positive quantity for p-channel devices fabricated on n-type substrates.

Using (2.44), values of $\phi_{MS'} + 2\phi_F$ obtained from Figures 2.9 and 2.11, and values of maximum surface depletion region charge density per unit area obtained from Figure 2.6, theoretical threshold voltages for *aluminum-gate n*-channel MOSFETs, operating at room temperature, can be calculated as a function of the acceptor doping concentration in the uniformly doped *p*-type silicon substrate, the magnitude of the fixed positive charge density per unit area located at the oxide-silicon interface, and the thickness of the silicon dioxide gate insulator.[4] The behavior of the theoretical threshold voltage for an *n*-channel aluminum-gate MOS transistor with a gate insulator consisting of 1000 A° of silicon dioxide as a function of Q_{SS} and substrate doping level can be seen in the curves shown in Figure 2.12.

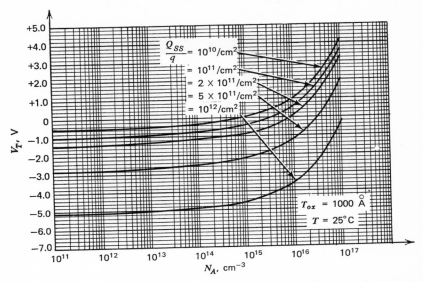

FIGURE 2.12 Theoretical values of threshold voltage for aluminum-gate n-channel MOSFETs with $T_{ox} = 1000\,\text{Å}$ at room temperature as a function of substrate doping concentration and fixed positive interface charge density.

Besides the characteristics associated with the active switching transistors in a MOS integrated circuit, the threshold voltages associated with aluminum interconnections passing over thick oxide layers which cross nonrelated diffused regions are also of prime importance. As is shown in Figure 2.13, if the voltage on the aluminum interconnect becomes too high, surface inversion will take place as the thick oxide region plays the role of a "gate insulating layer," and an *undesired* conduction path will form at the silicon surface between the two nonrelated diffusions. The resulting cross-coupling between active regions of the circuit could very well result in device failure and, for this reason, it is desirable to have the threshold voltages associated with the thick-oxide regions of an integrated circuit as *high* as possible, so that the active devices may be driven with high voltage levels without the possibility of parasitic conduction paths forming between neighboring transistors or diffused regions. This can be accomplished, as can readily be seen from (2.44), by increasing the thickness of the silicon dioxide in the regions between the individual transistors and diffused regions. However, as is seen in later chapters, there are basic processing problems which limit the maximum thickness of silicon dioxide which can be used. Typically, the thick oxide layers in MOS integrated circuits are on the order of 15,000 Å. Theoretical curves of the thick-field threshold voltage, V_{TT}, for n-channel aluminum-gate

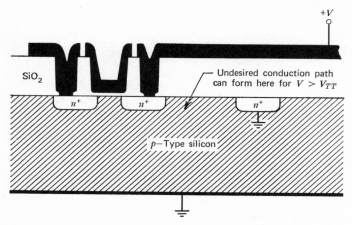

FIGURE 2.13 Parasitic transistor action in the thick-field region of an *n*-channel MOS integrated circuit.

thick oxide structures with this thickness of silicon dioxide are plotted in Figure 2.14 as a function of Q_{SS} and the doping concentration in the substrate. Since it is frequently necessary to relate the doping concentration in the silicon substrate to the resistivity in ohm-centimeters in the calculation of theoretical threshold voltages and other equally important processing parameters, the well-known experimental results reported on by Irvin[6] relating resistivity to doping concentration in both *p*-type and *n*-type silicon are presented in graphical form, for the reader's convenience, in Figure 2.15.

It can be seen from the curves of Figures 2.12 and 2.14 that whether an *n*-channel MOSFET is of the *depletion* or of the *enhancement* type will be a function of the acceptor doping concentration and the magnitude of Q_{SS}. The fact that it is more difficult to fabricate an *n*-channel enhancement device if the amount of fixed positive charge per unit area at the oxide-silicon interface is large is easily seen from the theoretical curves. For high substrate resistivities and low values of Q_{SS}, the threshold voltage is primarily determined by the term $\phi_{MS'} + 2\phi_F$, and will be slightly negative. For low values of Q_{SS} and low-resistivity substrates, the threshold voltage will be primarily determined by the maximum depletion region charge density per unit area, and enhancement-type operation with positive threshold voltages will be easily achievable. For high substrate resistivities, the effect of the charge in the depletion region will be almost negligible. Although the curves indicate that the observed threshold voltages of devices fabricated on uniformly doped silicon substrates with silicon dioxide gate insulators will always be negative for very low values of acceptor doping concentration in the substrate, the use of low-level diffusion techniques,[7] gold-doping,[8] ion-implantation,[9] or

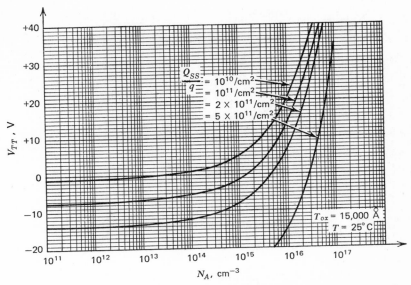

FIGURE 2.14 Theoretical values of thick-field threshold voltage for aluminum-gate n-channel thick-oxide structures with $T_{ox} = 15,000$ Å at room temperature as a function of substrate doping concentration and fixed positive interface charge density.

multilayer gate dielectrics[10] have been shown to be capable of achieving enhancement-type operation of n-channel MOS devices fabricated on high-resistivity substrates.

2.2.2 Threshold Voltages of p-channel Metal-Oxide-Silicon Devices

The charge distribution, under equilibrium conditions, for a p-channel MOSFET with a silicon dioxide gate insulator which has been fabricated on a uniformly doped n-type silicon substrate is shown in Figure 2.16. Figure 2.16a illustrates the situation when relatively small values of negative voltages are applied to the gate electrode and only a surface depletion region, consisting of nonmobile ionized donors extending to $x = x_d$, is present. Figure 2.16b illustrates the situation when the gate voltage is sufficiently negative to invert the surface of the n-type silicon and to form a p-type inversion layer. The notation used in Figure 2.16 is the same as was used in Figure 2.7, except that the charge density present in the (p-type) surface inversion layer per unit area is now denoted by Q_p. Both the charge densities present in the inversion layer and the depletion layer are positive quantities when the substrate material is n-type and are negative quantities when the

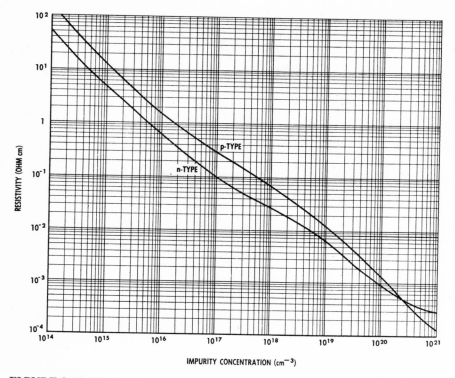

FIGURE 2.15   Resistivity of *p*- and *n*-type silicon as a function of impurity doping concentration at 300°K. (After Irvin[6]; reprinted with the permission of the American Telephone and Telegraph Company.)

substrate material is *p*-type. Since *p*-channel MOSFETs are almost always fabricated on *n*-type substrates, it follows that the positive inversion layer and depletion layer charge densities will add to the fixed positive charge density located at the oxide-silicon interface. Hence the charge present on the gate electrode and the threshold voltage of the device both will be negative as a result of the charge neutrality condition.

The threshold voltage of a *p*-channel MOSFET fabricated on a uniformly doped *n*-type silicon substrate will be given by (2.44), where the maximum value of the charge density per unit area present in the surface depletion region is approximately equal to

$$Q_{SD_{max}} \cong +(-4q\epsilon_s N_D \phi_F)^{1/2}. \qquad (2.45)$$

Since the quantities $-Q_{SS}$, $-Q_{SD_{max}}$, and $+(\phi_{MS'} + 2\phi_F)$ in the threshold

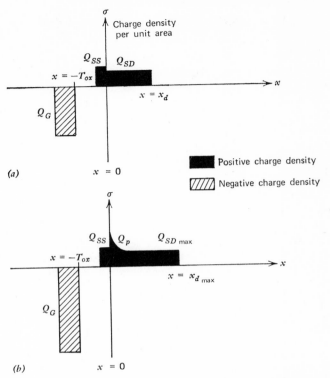

FIGURE 2.16   The charge distribution in a p-channel MOSFET (a) before inversion; (b) after inversion.

voltage equation are all negative, it follows that the threshold voltage of a p-channel device will always be negative, independent of the donor doping concentration in the silicon substrate or the value of Q_{SS}. However, in cases where p-channel depletion-mode MOS characteristics are desired, a number of techniques may be employed. The most widely used method to achieve p-channel depletion mode devices is to implant a narrow layer of acceptor-type impurities through the gate insulator and into the silicon surface between the drain and source regions, thereby providing a conducting p-type layer between the two regions even with zero gate voltage.[9]

Using (2.44), values of $\phi_{MS'} + 2\phi_F$ obtained from Figures 2.10 and 2.11, and values of maximum surface depletion region charge density per unit area obtained from Figure 2.6, theoretical threshold voltages for aluminum-gate p-channel MOSFETs operating at room temperature can be calculated as a function of the donor doping concentration in the uniformly doped n-type silicon substrate, the value of the fixed positive charge density per unit area located at the oxide-silicon interface, and the thickness of the silicon dioxide gate insulator, as was done before for n-channel devices. The behavior of the theoretical threshold voltage for p-channel aluminum-gate MOS transistors with gate insulators consisting of 1000 Å of silicon dioxide is shown in Figure 2.17 as a function of Q_{SS} and substrate doping level. Theoretical curves of the thick-field threshold voltages associated with p-channel aluminum-gate structures with 15,000 Å of silicon dioxide separating the aluminum from the underlying silicon substrate are shown in Figure 2.18. The behavior of the threshold voltage for p-channel MOSFETs as a function of the doping concentration in the substrate is very similar to the behavior of n-channel devices. For low values of Q_{SS}, the threshold voltage is primarily determined by the term $\phi_{MS'} + 2\phi_F$ for low donor doping concentrations in the substrate, and is primarily determined by the maximum charge density in the surface depletion region per unit area for high donor doping concentrations. As previously mentioned, for the case of a uniformly doped substrate and a gate insulator consisting only of silicon dioxide, p-channel devices will always exhibit negative threshold voltages. The threshold voltage will be more negative for lower resistivity substrates when all other parameters are held constant.

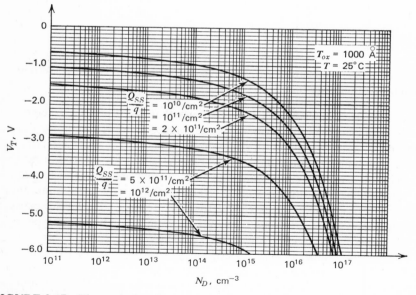

FIGURE 2.17 Theoretical values of threshold voltage for aluminum-gate p-channel MOSFETs with $T_{ox} = 1000$ Å at room temperature as a function of substrate doping concentration and fixed positive interface charge density.

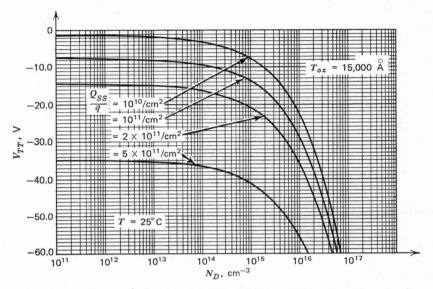

FIGURE 2.18   Theoretical values of thick-field threshold voltage for aluminum-gate *p*-channel thick-oxide structures with $T_{ox} = 15,000$ Å at room temperature as a function of substrate doping concentration and fixed positive interface charge density.

2.2.3   The Effect of Ionic Charge in the Gate Insulator on the Threshold Voltage of a MOS Transistor

In the derivation of the expression for the threshold voltage of a MOSFET, it was assumed that all of the charge in the gate oxide was positive, immobile, and located in very close proximity to the oxide-silicon interface. In practice, however, this is not always entirely true. Although the ionic charge density present in the gate oxide layer is usually positive, it can consist of both mobile and nonmobile species and can exist throughout the oxide layer. Unless extreme care is taken in the fabrication of MOS devices, mobile positive sodium ion contamination of the gate insulator can occur and the gradual movement of these ions under the influence of the applied gate voltage will result in long-term shifts of the observed threshold voltage. Nonmobile positive ionic charge can be generated within the oxide layer when a MOS device is exposed to radiation bombardment. These effects are treated in greater detail in later chapters. It is important to note that the presence of any ionic charge density in the gate insulator of a MOS device will have a pronounced effect on its threshold voltage. This can be easily seen by examining the simple MOS structure shown in Figure 2.19. If a charge sheet of

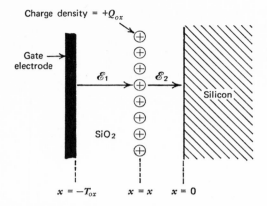

FIGURE 2.19 Ionic charge sheet within the gate oxide.

magnitude $+Q_{ox}$ per unit area is located a distance x from the oxide-silicon interface, a discontinuity in the normal component of the electric field will result across the charge sheet such that

$$\epsilon_{ox}(\mathscr{E}_2 - \mathscr{E}_1) = +Q_{ox}. \tag{2.46}$$

Also, if the voltage across the MOS structure is arbitrarily set equal to zero, then, with $-T_{ox} < x < 0$,

$$\mathscr{E}_1(T_{ox} + x) - \mathscr{E}_2 x = 0. \tag{2.47}$$

Equation 2.46 can be rewritten as

$$\mathscr{E}_1 = \mathscr{E}_2 - \frac{Q_{ox}}{\epsilon_{ox}}, \tag{2.48}$$

and substitution into (2.47) yields

$$\mathscr{E}_2 x = \mathscr{E}_2 T_{ox} + \mathscr{E}_2 x - \frac{Q_{ox} T_{ox}}{\epsilon_{ox}} - \frac{Q_{ox} x}{\epsilon_{ox}}, \tag{2.49}$$

or

$$\mathscr{E}_2 = + \frac{Q_{ox}}{\epsilon_{ox}}\left(1 + \frac{x}{T_{ox}}\right). \tag{2.50}$$

The $\mathscr{E}_2$ must be terminated in the silicon by a charge density per unit area of magnitude

$$Q_s = -\epsilon_{ox}\mathscr{E}_2 = -Q_{ox}\left(1 + \frac{x}{T_{ox}}\right). \tag{2.51}$$

It is apparent that when the charge sheet is located at $x \simeq 0$, the oxide charge induces an equal and opposite amount of charge in the silicon and, consequently, acts in the same manner as the fixed positive interface charge density, Q_{SS}. However, when the oxide charge is located at $x = -T_{ox}$, the charge density induced at the silicon surface will be equal to zero. Hence the amount of charge that will be induced in the silicon will depend on how close the oxide charge sheet is to the oxide-silicon interface. The effect of a single oxide charge sheet on the threshold voltage of a MOS device can be considered to be that of an "effective" interface charge density in addition to the actual value of Q_{SS}. That is,

$$V_T = \left(\frac{-Q_{SS} - Q_{\text{eff}} - Q_{SD\text{max}}}{\epsilon_{ox}}\right)T_{ox} + \phi_{MS'} + 2\phi_F, \qquad (2.52)$$

where Q_{eff} is given by

$$Q_{\text{eff}} \equiv Q_{ox}\left(1 + \frac{x}{T_{ox}}\right) = -Q_s. \qquad (2.53)$$

For an arbitrary ionic charge distribution in the oxide that is a function of x only, the effect on the threshold voltage of the device can be obtained by considering the charge distribution as an infinite sum of individual charge sheets spaced within the oxide. In terms of $\rho(x)$, the charge density per unit volume in the oxide layer, the result is obtained by integration:

$$V_T = \left\{\frac{-Q_{SS} - \int_{x=-T_{ox}}^{x=0-} [1 + (x/T_{ox})]\rho(x)\,dx - Q_{SD\text{max}}}{\epsilon_{ox}}\right\}T_{ox} + \phi_{MS'} + 2\phi_F.$$

$$(2.54)$$

The upper limit of the integration is taken to be $x = 0-$; thus the actual fixed positive interface charge density per unit area will not contribute to the integral.

Charge neutrality is still preserved in the MOS structure when ionic charges are present within the oxide. While the nth individual charge sheet will induce a charge density per unit area $-Q_{sn}$ in the silicon, an additional charge density of magnitude $-Q_{oxn} + Q_{sn}$ is also induced on the gate, where Q_{oxn} is the magnitude of the charge density per unit area on the nth charge sheet. The charge neutrality condition at the onset of strong inversion is now:

$$Q_G + Q_{SS} + Q_{SD\text{max}} + \int_{x=-T_{ox}}^{x=0-} \rho(x)\,dx = 0. \qquad (2.55)$$

Because the ionic charge in the oxide has more effect on the threshold voltage when it is closer to the oxide-silicon interface, the drift of charged mobile ions through the gate oxide layer under the influence of a high electric field results in an instability in the characteristics of MOS field-effect devices. For example, if the gate oxide layer is contaminated with positively charged sodium ions, when a large positive gate voltage is applied, the ions will drift toward the oxide-silicon interface, where will have the maximum effect on the threshold voltage and the threshold voltage becomes more negative. On the other hand, when a large negative gate voltage is applied, the sodium ions will be attracted back toward the gate electrode where they will have practically no effect on the characteristics of the surface space charge region and the threshold voltage will tend to shift back in the positive direction.

2.3 THE VARIATION OF THE THRESHOLD VOLTAGE OF MOS TRANSISTORS WITH APPLIED SUBSTRATE-TO-SOURCE BIAS

Although MOSFETs are frequently operated with their source and substrate regions connected and held at zero potential, the consideration of the characteristics of these devices under the influence of an applied substrate-to-source voltage is also important. The effects of any potential difference between the source and substrate regions on the electrical parameters of MOSFETs are of particular interest because such situations are commonly encountered in integrated circuits, where many of the individual devices have source regions that are not connected directly to ground and whose potential will vary with respect to the substrate during normal circuit operation. Figure 2.20 illustrates an n-channel MOSFET that is operating with an

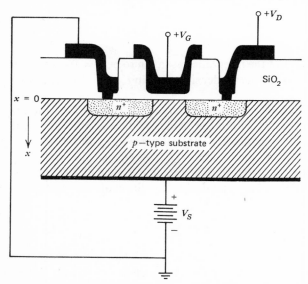

FIGURE 2.20 n-Channel MOSFET operating with applied substrate-to-source voltage.

applied substrate-to-source voltage, V_S. For the case of an n-channel device, the voltage applied to the substrate must be *negative* with respect to the source so that all junctions to substrate remain reverse-biased. For p-channel devices, on the other hand, the applied substrate-to-source voltage should be *positive*.

In general, the presence of a potential difference between the substrate and source regions of a MOSFET will have very little effect on the gain of the device, but will greatly influence the characteristics of the surface space charge region and the threshold voltage. In addition, the reverse substrate bias will tend to decrease the junction capacitances of all diffused regions to substrate. The behavior of the surface space charge region and the threshold voltage of a MOSFET with applied substrate-to-source voltage can best be examined by once again considering Poisson's equation, given by (2.1). As before, the electric field in the x direction and the space charge density can be described by (2.2) and (2.3), respectively. If the gate-to-source voltage is sufficiently high such that an inversion layer forms at the silicon surface, the application of a small negative voltage to the p-type substrate will tend to reverse-bias the junction formed between the surface inversion layer (which, being in ohmic contact with the grounded source region, is kept at essentially zero potential) and the substrate. Consequently, the negative substrate bias will widen the surface depletion region beyond its maximum equilibrium value, $x_{d_{max}}$, as shown in Figure 2.21.[11] As is also shown in the figure, the surface potential ϕ_s, which is a direct measure of the amount of surface band-bending, also increases with increasingly negative substrate voltage.

Since the application of the substrate-to-source voltage establishes a nonequilibrium situation, the use of quasi-Fermi levels is appropriate. The quasi-Fermi potential for the majority carriers in the substrate is equal to the equilibrium Fermi potential ϕ_{Fp} and does not vary with position. The quasi-Fermi potential for the electrons, which are minority carriers in the p-type substrate, increases with negative substrate voltage, as shown in Figure 2.21, and is equal to the sum of the equilibrium Fermi potential plus the applied substrate voltage (with respect to the source). Within the surface space charge region, the quasi-Fermi potentials can be written as

$$\phi_{Fn} = \phi_F - V_S$$

and $\quad\quad\quad\quad\quad\quad\quad\quad\quad\quad\quad\quad\quad\quad\quad\quad\quad\quad\quad\quad$ (2.56)

$$\phi_{Fp} = \phi_F.$$

Deep within the bulk of the substrate, $\phi_{Fn} = \phi_{Fp} = \phi_F$ and the hole and electron concentrations per unit volume are as given in (2.5). As before, charge neutrality must exist within the bulk of the substrate. Assuming that all impurities are ionized results in the condition that

$$N_D - N_A = n - p,$$ $\quad\quad\quad\quad\quad\quad$ (2.57)

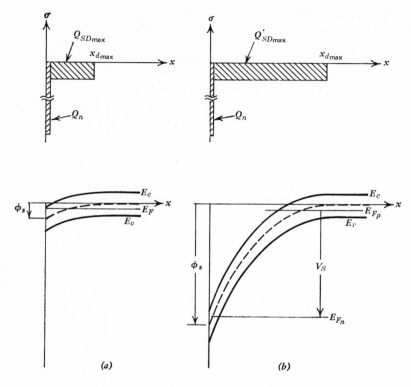

FIGURE 2.21   Idealized charge distribution and energy band diagram for an inverted p-type substrate in a direction normal to the surface, some distance away from an n^+ diffused region; (a) equilibrium case with $V_S = 0$ V; (b) reverse-bias case with $V_S = -5$ V. (After Grove and Fitzgerald.[11])

where the quantity $N_D - N_A$ is given by (2.6). In terms of the applied substrate-to-source voltage V_S, the carrier concentrations in the surface space charge region are equal to

$$n = n_i \exp \left[\frac{-q(\phi_F - V_S - \phi)}{kT} \right]$$

and

$$p = n_i \exp \left[\frac{+q(\phi_F - \phi)}{kT} \right].$$

(2.58)

Therefore, from (2.1), (2.3), (2.6), and (2.58), Poisson's equation can be rewritten as

$$\frac{\partial^2 \phi}{\partial x^2} = -\frac{q n_i}{\epsilon_s}\left[2 \sinh\left(\frac{-q\phi_F}{kT}\right) + \exp\left(\frac{+q(\phi_F - \phi)}{kT}\right)\right.$$
$$\left. - \exp\left(\frac{-q(\phi_F - V_S - \phi)}{kT}\right)\right]. \quad (2.59)$$

Once again, the notation can be simplified by defining the following quantities:

$$u \equiv \frac{q\phi}{kT}$$

$$u_F \equiv \frac{q\phi_F}{kT}$$

$$u_s \equiv \frac{q\phi_s}{kT} \quad\quad (2.60)$$

$$v_S \equiv \frac{qV_S}{kT},$$

and since $\sinh(-u_F) = -\sinh(u_F)$, then (2.59) can be written as

$$\frac{\partial^2 u}{\partial x^2} = +\frac{q^2 n_i}{\epsilon_s kT}\left\{2 \sinh(u_F) - \exp(u_F - u) + \exp\left[-(u_F - v_S - u)\right]\right\}.$$
$$(2.61)$$

Equation 2.61 can be expressed in terms of the intrinsic Debye length L_D as

$$\frac{\partial^2 u}{\partial x^2} = \frac{1}{L_D{}^2}\left\{\sinh(u_F) - \tfrac{1}{2}\exp(u_F - u) + \tfrac{1}{2}\exp\left[-(u_F - v_S - u)\right]\right\}. \quad (2.62)$$

Integration from the bulk (where $\partial u/\partial x = 0$) toward the surface gives

$$\int_0^{\partial u/\partial x}\left(\frac{\partial u}{\partial x}\right) d\left(\frac{\partial u}{\partial x}\right)$$
$$= \frac{1}{L_D{}^2}\int_0^u \left\{\sinh(u_F) - \tfrac{1}{2}\exp(u_F - u) + \tfrac{1}{2}\exp\left[-(u_F - v_S - u)\right]\right\} du,$$
$$(2.63)$$

or

$$\frac{1}{2}\left(\frac{\partial u}{\partial x}\right)^2 = \frac{1}{L_D{}^2}\left[u \sinh(u_F) + \frac{e^{u_F}}{2}(e^{-u} - 1) + \frac{e^{(-u_F + v_S)}}{2}(e^{+u} - 1)\right]. \quad (2.64)$$

The electric field in the x direction can be expressed as $\mathscr{E}_x = -(\partial\phi/\partial x) = -(kT/q)(\partial u/\partial x)$, and by choosing the negative root to obtain the proper direction for the field intensity, one obtains

$$\mathscr{E}_x = +\frac{kT}{q}\left\{\frac{2}{L_D^2}\left\{u\,\sinh\,(u_F)\right.\right.$$
$$\left.\left.+\left[\frac{e^{(u_F-u)} - e^{(u_F)} + e^{(-u_F+v_S+u)} - e^{(-u_F+v_S)}}{2}\right]\right\}\right\}^{1/2}. \quad (2.65)$$

At the silicon surface, $x = 0$ and $u = u_s$ and the electric field is equal to

$$\mathscr{E}_s = \frac{kT}{qL_D}\,[2u_s\,\sinh\,(u_F) + e^{(u_F-u_S)} - e^{(u_F)} + e^{(-u_F+v_S+u_S)} - e^{(-u_F+v_S)}]^{1/2}. \quad (2.66)$$

The total charge density per unit area in the silicon required to terminate this electric field is given by

$$Q_{\text{total}} = -\epsilon_s\mathscr{E}_s, \quad (2.67)$$

where Q_{total} is equal to the sum of Q_n, the charge density per unit area in the n-type inversion layer, and $Q'_{SD\text{max}}$, the charge density present in the surface depletion region per unit area with applied substrate-to-source voltage. Combining (2.66) and (2.67), the total charge density at the surface of the p-type substrate can be expressed as

$$Q_{\text{total}} = -2qn_iL_D[2u_s\,\sinh\,(u_F) + e^{(u_F-u_S)} - e^{(u_F)}$$
$$+ e^{(-u_F+v_S+u_S)} - e^{(-u_F+v_S)}]^{1/2}, \quad (2.68)$$

where L_D, the intrinsic Debye length, is given by (2.12). With an applied substrate voltage, the surface potential under conditions of strong inversion is approximately

$$\phi_s \cong 2\phi_F - V_S, \quad (2.69)$$

or

$$u_s \cong 2u_F - v_S. \quad (2.70)$$

Once again assuming that when the applied gate-to-source potential becomes sufficiently large for the onset of strong inversion, Q_n is still approximately equal to zero and

$$Q_{\text{total}}\big|_{\phi_s \cong 2\phi_F - V_S} \cong Q'_{SD\text{max}}, \quad (2.71)$$

then the maximum value of surface depletion region charge density per unit area with applied substrate-to-source voltage can be obtained through (2.68), (2.70), and (2.71). The result is simply

$$Q'_{SD\text{max}} \cong -2qn_iL_D[2(2u_F - v_S)\,\sinh\,(u_F)]^{1/2}. \quad (2.72)$$

For temperatures and substrate doping concentrations such that $u_F \gg 1$, (2.72) can be approximated by

$$Q'_{SD_{max}} \cong -2qn_iL_D[(2u_F - v_S)e^{(u_F)}]^{1/2}, \tag{2.73}$$

or

$$Q'_{SD_{max}} \cong -(2n_ikT\epsilon_se^{(u_F)})^{1/2}[(2u_F - v_S)]^{1/2}. \tag{2.74}$$

Now, since $N_A \cong n_ie^{(u_F)}$, (2.74) reduces to

$$Q'_{SD_{max}} \cong -[2\epsilon_skTN_A(2u_F - v_S)]^{1/2}, \tag{2.75}$$

or

$$Q'_{SD_{max}} \cong -[2\epsilon_sN_Aq(2\phi_F - V_S)]^{1/2}. \tag{2.76}$$

The ratio of the maximum value of surface depletion charge density per unit area with an applied substrate-to-source voltage of V_S to the maximum value with zero substrate-to-source voltage can be obtained by dividing (2.76) by (2.29). The result is

$$\frac{Q'_{SD_{max}}}{Q_{SD_{max}}} \cong \left[\frac{2\phi_F - V_S}{2\phi_F}\right]^{1/2}, \tag{2.77}$$

or

$$Q'_{SD_{max}} \cong Q_{SD_{max}}\left[\frac{2\phi_F - V_S}{2\phi_F}\right]^{1/2} \tag{2.78}$$

For large values of applied substrate-to-source voltage, both the maximum value of the charge density per unit area in the surface depletion region, $Q'_{SD_{max}}$, and the maximum width of the surface depletion region, $x_{d_{max}}$, will increase substantially over the values obtained for these quantities with $V_S = 0$ V. The threshold voltage associated with an n-channel MOSFET will become increasingly positive as the applied (negative) substrate-to-source potential becomes larger. This is a direct result of the increase in $Q'_{SD_{max}}$ according to (2.78). Combining (2.44) and (2.78) yields the expression for the threshold voltage of an n-channel MOSFET operating with applied substrate-to-source bias:

$$V_T \cong \left\{\frac{-Q_{SS} - Q_{SD_{max}}[(2\phi_F - V_S)/2\phi_F]^{1/2}}{\epsilon_{ox}}\right\}T_{ox} + \phi_{MS'} + 2\phi_F. \tag{2.79}$$

The *change* in threshold voltage as a function of the applied substrate-to-source voltage is simply

$$\Delta V_T \cong \frac{-Q_{SDmax}T_{ox}}{\epsilon_{ox}}\left[\left(\frac{2\phi_F - V_S}{2\phi_F}\right)^{1/2} - 1\right]. \tag{2.80}$$

Note that since $Q_{SD_{max}}$ is a negative quantity for an n-channel device fabricated on a p-type substrate, ΔV_T will be positive for negative V_S.

By following the same procedure that was used to derive an expression for the functional dependence of the surface depletion region charge density per unit area with respect to the applied substrate voltage for n-channel MOSFETs, $Q'_{SD_{max}}{}'$ for p-channel devices fabricated on n-type silicon substrates

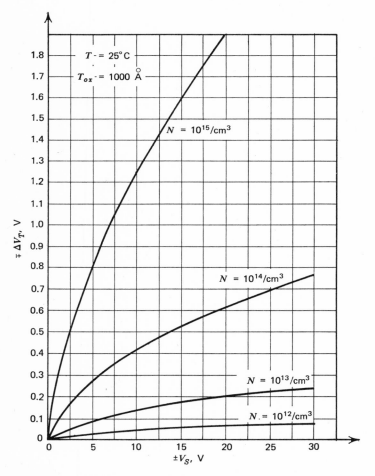

FIGURE 2.22   Variation of threshold voltage with applied substrate-to-source bias per 1000 Å of silicon dioxide gate insulator for n- and p-channel MOSFETs at room temperature, as a function of substrate doping concentrations between 10^{12} and 10^{15}/cm^3.

can easily be shown to be approximately

$$Q'_{SD_{max}} \cong +[-2qN_D\epsilon_s(2\phi_F - V_S)]^{1/2}, \qquad (2.81)$$

under conditions such that $u_F < -1$ and $\exp(u_F) \ll 1$. Now V_S will, of course, be *positive* for reverse bias of p-channel devices. The ratio of the maximum value of the surface depletion region charge density per unit area

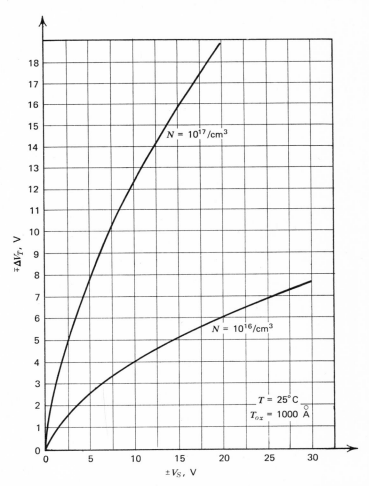

FIGURE 2.23   Variation of threshold voltage with applied substrate-to-source bias per 1000 Å of silicon dioxide gate insulator for n- and p-channel MOSFETs at room temperature, as a function of substrate doping concentrations between 10^{16} and $10^{17}/cm^3$.

with applied substrate bias to its value with zero bias will be

$$Q'_{SD_{\max}} \cong Q_{SD_{\max}} \left[\frac{-2\phi_F + V_S}{-2\phi_F} \right]^{1/2} \tag{2.82}$$

and, consequently, the threshold voltage of a p-channel MOSFET operating with positive substrate-to-source bias will be given by

$$V_T \cong \left\{ \frac{-Q_{SS} - Q_{SD_{\max}}[(-2\phi_F + V_S)/-2\phi_F]^{1/2}}{\epsilon_{ox}} \right\} T_{ox} + \phi_{MS'} + 2\phi_F. \tag{2.83}$$

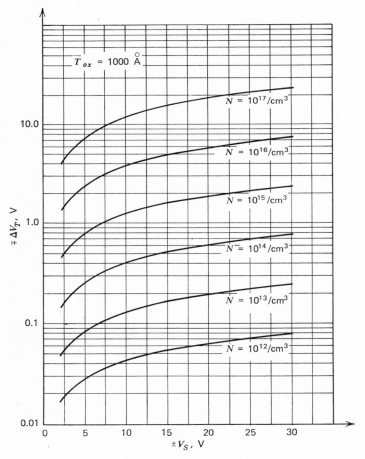

FIGURE 2.24   Semilogarithmic representation of the data from Figures 2.23 and 2.24, showing $\Delta V_T(V_S)$ for doping concentrations from 10^{12} to $10^{17}/\text{cm}^3$.

The *change* in threshold voltage as a function of V_S is given by

$$\Delta V_T = -\frac{Q_{SD_{max}} T_{ox}}{\epsilon_{ox}} \{[(-2\phi_F + V_S)/-2\phi_F]^{1/2} - 1\}. \qquad (2.84)$$

Since $Q_{SD_{max}}$ is a positive quantity for a *p*-channel device fabricated on an *n*-type substrate, it follows that ΔV_T will be negative for positive substrate-to-source voltage.

Graphs of the change in threshold voltage of a MOSFET as a function of different substrate doping concentrations are shown in Figures 2.22 and 2.23 for room temperature operation. Since ΔV_T is directly proportional to the thickness of the gate insulator, the graphs have been normalized to 1000 Å of silicon dioxide for the gate insulating layer. The values of ΔV_T shown in the figures are a result of numerical evaluation of the theoretical equations presented above in conjunction with values of $Q_{SD_{max}}$ and ϕ_F which were previously evaluated in the text. It should be noted that Figures 2.22 and 2.23 are valid for *both* *n*-channel and *p*-channel devices, with the appropriate choice of sign. For convenience, the data from Figures 2.22 and 2.23 have been condensed into the semilogarithmic representation shown in Figure 2.24. As was previously mentioned, although the application of substrate-to-source

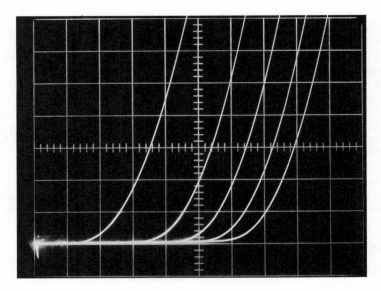

FIGURE 2.25 Room temperature operating characteristics of an *n*-channel MOSFET. Vertical scale (drain current): +1 mA/div.; horizontal scale (gate-to-source voltage): +5 V/div.; drain voltage = +20 V constant; V_S = 0, −5, −10, −15, and −20 V.

bias can have a substantial effect on the threshold voltage of a MOSFET, it usually will have very little effect on the gain of the device. This can easily be seen in Figure 2.25, which illustrates the change in the operating characteristics of an n-channel MOS transistor with $V_S = 0$, -5, -10, -15, and -20 V.

REFERENCES

1. R. H. Kingston and S. F. Neustadter, Calculation of the Space Charge, Electric Field, and Free Carrier Concentration at the Surface of a Semiconductor, *Journal of Applied Physics*, Vol. 26, No. 6, 1955, pp. 718–720.

2. A. S. Grove, B. E. Deal, E. H. Snow, and C. T. Sah, Investigation of Thermally Oxidized Silicon Surfaces Using Metal-Oxide-Semiconductor Structures, *Solid State Electronics*, Vol. 8, 1965, pp. 145–163.

3. A. S. Grove, *Physics and Technology of Semiconductor Devices*, John Wiley and Sons, New York, 1967, p. 268.

4. P. Richman, Theoretical Threshold Voltages for MOS Field-Effect Transistors, *Solid State Electronics*, Vol. 11, 1968, pp. 869–876.

5. B. E. Deal, E. H. Snow, and C. A. Mead, Barrier Energies in Metal-Silicon Dioxide-Silicon Structures, *Journal of Physics and Chemistry of Solids*, Vol. 27, 1966, pp. 1873–1879.

6. J. C. Irvin, Resistivity of Bulk Silicon and of Diffused Layers in Silicon, *Bell System Technical Journal*, Vol. 41, 1962, p. 387.

7. T. Athanas, Development of Low Threshold Voltage Complementary MOS Integrated Circuits, *Proceedings of the 1971 IEEE Convention*, New York, March 1971, pp. 538–539.

8. P. Richman, The Effect of Gold Doping upon the Characteristics of MOS Field-Effect Transistors with Applied Substrate Voltage, *Proceedings of the IEEE*, Vol. 56, No. 4, 1968, pp. 774–775.

9. K. G. Aubuchon, The Use of Ion Implantation to Set the Threshold Voltage of MOS Transistors, presented at the International Conference on Properties and Use of M.I.S. Structures, Grenoble, France, June 17–20, 1969.

10. H. E. Nigh, J. Stach, and R. M. Jacobs, A Sealed Gate IGFET, presented at the 1967 Solid State Device Research Conference, Santa Barbara, California.

11. A. S. Grove and D. J. Fitzgerald, Surface Effects on *p-n* Junctions: Characteristics of Surface Space Charge Regions Under Non-Equilibrium Conditions, *Solid State Electronics*, Vol. 9, 1966, pp. 783–806.

BIBLIOGRAPHY

Brotherton, S. D., Dependence of MOS Transistor Threshold Voltage on Substrate Resistivity, *Solid State Electronics*, Vol. 10, 1967, pp. 611–616.

Cagnina, S. F., and E. H. Snow, Properties of Gold-Doped MOS Structures, *Journal of the Electrochemical Society*, Vol. 114, No. 11, November 1967, pp. 1165–1172.

Chang, C. Y., and K. Y. Tsao, Electrical Properties of Diffused Zinc on SiO_2-Si MOS Structures, *Solid State Electronics*, Vol. 12, 1969, pp. 411–415.

Collins, D. R., The Effect of Gold on the Properties of the Si-SiO_2 System, *Journal of Applied Physics*, Vol. 39, No. 9, August 1968, pp. 4133–4143.

Das, M. B., Dependence of the Characteristics of MOS Transistors on the Substrate Resistivity, *Solid State Electronics*, Vol. 11, 1968, pp. 305–322.

Dillon, Jr., J. A., and H. E. Farnsworth, Work Function and Sorption Properties of Silicon Crystals, *Journal of Applied Physics*, Vol. 29, No. 8, August 1958, pp. 1195–1202.

Frankl, D. R., Conditions for Quasi-Equilibrium in a Semiconductor Surface Space-Charge Layer, *Surface Science*, Vol. 3, 1965, pp. 101–108.

Gobeli, G. W., and F. G. Allen, Direct and Indirect Excitation Processes in Photoelectric Emission from Silicon, *Physical Review*, Vol. 127, No. 1, July 1962, pp. 141–149.

Goodman, A. M., Photoemission of Electrons from Silicon and Gold into Silicon Dioxide, *Physical Review*, Vol. 144, No. 2, April 1966, pp. 588–593.

Lindmayer, J., Field Effect Studies of the Oxidized Silicon Surface, *Solid State Electronics*, Vol. 9, 1966, pp. 225–235.

Lindmayer, J., Heterojunction Properties of the Oxidized Semiconductor, *Solid State Electronics*, Vol. 8, 1965, pp. 523–528.

Morin, F. J., and J. P. Maita, Electrical Properties of Silicon Containing Arsenic and Boron, *Physical Review*, Vol. 96, No. 1, October 1954, pp. 28–35.

Nassibian, A. G., Effect of Diffused Oxygen and Gold on Surface Properties of Oxidized Silicon, *Solid State Electronics*, Vol. 10, 1967, pp. 879–890.

Nassibian, A. G., Effect of Gold on Surface Properties and Leakage Current of MOS Transistors, *Solid State Electronics*, Vol. 10, 1967, pp. 891–896.

Sah, C. T., Characteristics of the Metal-Oxide-Semiconductor Transistors, *IEEE Transactions on Electron Devices*, Vol. ED-11, July, 1964, pp. 324–345.

Sah, C. T., and H. C. Pao, The Effects of Fixed Bulk Charge on the Characteristics of Metal-Oxide-Semiconductor Transistors, *IEEE Transactions on Electron Devices*, Vol. ED-13, No. 4, April 1966, pp. 393–409.

Van Nielen, J. A., and O. W. Memelink, The Influence of the Substrate Upon the DC Characteristics of Silicon MOS Transistors, *Philips Research Reports*, Vol. 22, 1967, pp. 55–71.

Wedlock, B. D., Direct Determination of the Pinch-Off Voltage of a Depletion-Mode Field-Effect Transistor, *Proceedings of the IEEE*, Vol. 57, No. 1, January 1969, pp. 75–77.

Wilder, E. M., Source Bias and the Large-Signal MOS Transistor Switch, *Electro-Technology*, September 1967, pp. 46–48.

Young, C. E., Extended Curves of the Space Charge, Electric Field, and Free Carrier Concentration at the Surface of a Semiconductor, and Curves of the Electrostatic Potential Inside a Semiconductor, *Journal of Applied Physics*, Vol. 32, No. 3, March 1961, pp. 329–332.

PROBLEMS

2.1 Derive an expression for the gate voltage required to cause sufficient band-bending at the silicon surface of an *n*-channel MOS transistor to make the surface intrinsic (i.e., $\phi_s = \phi_F$). Express the result in terms of Q_{SS}, $Q_{SD_{\max}}$, ϵ_{ox}, T_{ox}, $\phi_{MS'}$, and ϕ_F.

2.2 Consider a MOS transistor with a gate insulator consisting of a thermally grown layer of silicon dioxide covered by an overlying layer of silicon nitride. Assuming the gate insulator to be stable and free from charge transport mechanisms, show that the threshold voltage of the MOSFET will be given by

$$V_T = \left(\frac{-Q_{SS} - Q_{SD_{\max}}}{\epsilon_{ox}}\right)\left[T_{ox} + T_{\mathrm{Si_3N_4}}\left[\frac{\epsilon_{ox}}{\epsilon_{\mathrm{Si_3N_4}}}\right]\right] + \phi_{MS'} + 2\phi_F,$$

where $T_{\mathrm{Si_3N_4}}$ is the thickness of the layer of silicon nitride and $\epsilon_{\mathrm{Si_3N_4}}$ is the dielectric constant of the silicon nitride.

2.3 If the oxide layer in the metal-nitride-oxide-silicon (MNOS) transistor structure of problem 2.2 is made extremely thin (on the order of 10 to 50 A°) with the application of a sufficiently high positive gate-to-source voltage, electrons can tunnel from the silicon substrate across the oxide layer and become trapped in localized states at the oxide-nitride interface. Assuming that once the gate voltage is reset to zero, a charge density $-Q$ now resides at the oxide-nitride interface, what is the resulting change in the threshold voltage of the device? Comment on what effect this might have on the characteristics of a p-channel enhancement-type MOSFET.

2.4 Aluminum-gate p-channel MOS transistors with three different thicknesses of silicon dioxide gate insulator (T_{ox1}, T_{ox2}, and T_{ox3}) are fabricated on the same n-type silicon substrate. The donor doping concentration in the substrate is equal to N_D. The observed values of threshold voltage for the individual transistors are V_{T1}, V_{T2}, and V_{T3}, respectively. When the threshold voltages are plotted versus the thickness of the silicon dioxide gate insulator, the result is a straight line as shown in Figure 2.26. Assuming that the fixed

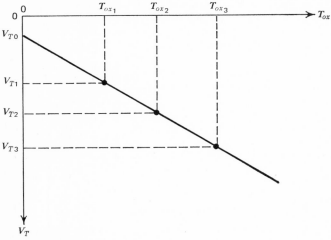

FIGURE 2.26

positive interface charge density, Q_{SS}, is uniform over the surface of the wafer: (a) Derive an expression for $Q_{SD_{max}}$ in terms of Q_{SS}, the observed threshold voltages and thicknesses associated with the devices, and the dielectric constant of the silicon dioxide. (b) Derive an expression for Q_{SS} in terms of the observed threshold voltages and oxide thicknesses associated with the individual devices, the dielectric constant of silicon dioxide, and the effective donor doping concentration in the n-type substrate. (c) Derive an expression for $\phi_{MS'}$ in terms of V_{T0} and the parameters above.

2.5 Most n-channel depletion units are usually fabricated by employing a conventional MOSFET structure on a high-resistivity silicon substrate. Such a device will normally exhibit very little change of threshold voltage with applied substrate-to-source reverse bias (source-body effect). How would one fabricate an n-channel depletion-type device with a threshold voltage that changed rapidly with increasing substrate-to-source reverse bias?

2.6 Consider the case of an n-channel depletion-type MOSFET that is "normally on" with zero gate-to-source voltage because of the presence of an n-type diffused or epitaxially grown surface layer extending a distance x_j into the p-type silicon substrate. For such a device, (a) would the drain-to-source conductance always be able to be cut off with negative gate-to-source voltage? (b) What is the constraint on the diffusion depth x_j, the doping gradient, and surface concentration? (compare x_j with $x_{d_{max}}$). (c) If the device could not be cut off with negative gate-to-source voltage, how else could it be cut off?

3

The Capacitance of a MOS Structure as a Function of Voltage and Frequency

A cross-sectional view of a simple MOS capacitor structure is illustrated in Figure 3.1. The structure consists of a silicon substrate, which can be either p- or n-type, covered by a thin insulating layer of silicon dioxide, and a metal gate electrode directly above. Unlike the MOS transistor, the MOS capacitor is a two-terminal device whose capacitance will vary with the applied gate-to-substrate voltage. Under d.c. conditions, no current can flow from gate to substrate because of the presence of the silicon dioxide insulator and, as a result, the capacitor is in thermal equilibrium and can be treated accordingly. The capacitance versus voltage characteristics of MOS capacitors that result from the modulation of the width of the surface space charge region by the gate field have been found to be extremely useful in the evaluation of the electrical properties of insulator-semiconductor interfaces. The structure is also frequently used to achieve capacitors for integrated circuit applications.

3.1 THE MOS CAPACITOR: QUALITATIVE DESCRIPTION OF OPERATION

There are three regions of interest when the capacitance of the MOS capacitor is plotted as a function of voltage. The case of a MOS capacitor fabricated on a p-type substrate is treated here. The characteristics of a similar device fabricated on an n-type substrate can be treated in an analogous manner.

47

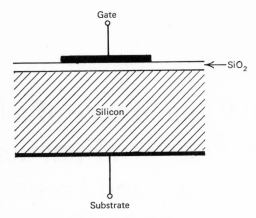

FIGURE 3.1   Cross-sectional view of a MOS capacitor.

3.1.1   Accumulation

For sufficiently large negative gate voltages, majority carrier holes will be attracted to the surface of the silicon to terminate the electric field in the gate insulator; consequently, a p-type surface accumulation layer will be formed in the silicon, as shown in Figure 3.2a. Because of the increase in the concentration of holes due to the applied gate voltage, the Fermi level near the silicon surface will move to a position closer to the valence band edge. Since the structure is in thermal equilibrium, the Fermi level remains constant with position and this results in the surface band-bending shown in Figure 3.2b.

The high concentration of holes near the accumulated silicon surface can be thought of as forming the second electrode of a parallel-plate capacitor with the gate electrode. Since the accumulation layer is in direct ohmic contact with the p-type substrate, the capacitance of the structure under accumulation conditions must be approximately equal to $\epsilon_{ox} A/T_{ox}$, where A is the area of the gate electrode.

3.1.2   Depletion

If the magnitude of the applied negative gate voltage is decreased, the hole concentration at the surface of the silicon will also decrease. As this process continues, the gate voltage can eventually be reduced to the point where the surface hole concentration will go to zero and only a surface depletion region consisting of nonmobile ionized acceptors will be required to terminate the electric field in the gate insulator, as is illustrated in Figure 3.3a. It can easily be seen that under depletion conditions the Fermi level near the silicon

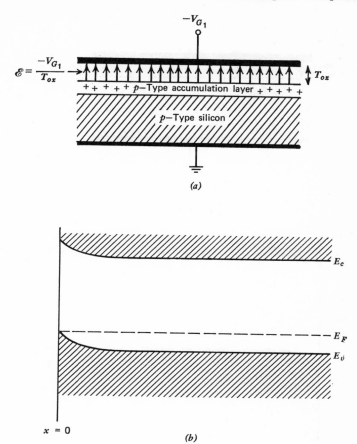

FIGURE 3.2   Surface accumulation in a MOS capacitor fabricated on a p-type silicon substrate.

surface will move to a position closer to the center of the forbidden region. This is shown in the energy band diagram of Figure 3b, which illustrates the nature of the surface band-bending when only a depletion region is present.

Since the magnitude of the charge density per unit area in the surface depletion region will be equal to the acceptor doping concentration times the electronic charge times the width of the surface depletion region, increasingly positive gate-to-substrate voltage will tend to increase both Q_{SD} and x_d. As the width of the surface depletion region increases, the capacitance from gate to substrate associated with the MOS capacitor structure will decrease, because the capacitance associated with the surface depletion

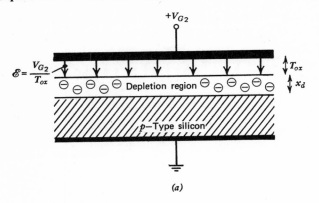

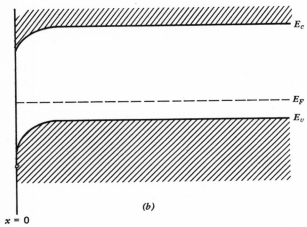

FIGURE 3.3 Surface depletion in a MOS capacitor fabricated on a p-type silicon substrate.

region will add in series to the capacitance across the gate insulator. Thus the total capacitance per unit area from gate to substrate under depletion conditions is given by

$$C(V_G) = \left(\frac{1}{C_{ox}} + \frac{1}{C_{SD}(V_G)}\right)^{-1}, \tag{3.1}$$

where C_{ox} is the oxide capacitance per unit area, ϵ_{ox}/T_{ox}, and C_{SD} is the capacitance per unit area associated with the surface depletion region. Then C_{SD} can easily be shown to be

$$C_{SD} = \frac{\epsilon_s}{x_d}, \tag{3.2}$$

where ϵ_s is the dielectric constant of silicon and x_d is the width of the surface depletion region, which will be a function of the applied gate voltage. It should be noted that (3.1) will be valid only when the amount of charge that might be trapped in surface states at the silicon-silicon dioxide interface is independent of the surface potential.

3.1.3 Inversion

With increasingly positive applied gate voltage, the surface depletion region will continue to widen until the onset of surface inversion is observed as

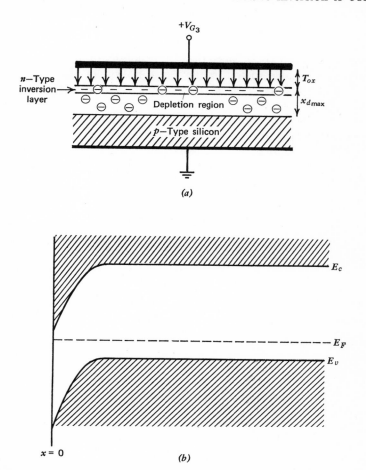

FIGURE 3.4 Surface inversion in a MOS capacitor fabricated on a p-type silicon substrate.

conduction band electrons are attracted up to the silicon surface to form an *n*-type inversion layer. This situation is illustrated in Figure 3.4*a*. Since the surface of the silicon has become strongly inverted because of the action of the gate field, the Fermi level near the silicon surface will now lie close to the conduction band edge, resulting in the surface band-bending shown in Figure 3.4*b*.

As was previously shown in Section 2.1, the width of the surface depletion region for a MOS structure in equilibrium will remain virtually constant after the formation of a surface inversion layer, even if the gate voltage is made more positive. Small variations in the width of the surface depletion region around its maximum value can occur, however, if a nonequilibrium situation exists where the charge density in the inversion layer is unable to follow a high-frequency small-signal a.c. voltage applied to the gate electrode, superimposed on the d.c. bias. Since the charge density in the inversion layer may or may not be able to follow the a.c. variation of the applied gate voltage, it follows that *the capacitance under inversion conditions will be a function of frequency*. In general, when a surface inversion layer is present, the gate-to-substrate capacitance of the MOS capacitor structure, for low-frequency a.c. signals, will be equal to the dielectric capacitance $C_{ox}A$. For high-frequency signals, the observed capacitance will be equal to the series combination of the dielectric capacitance and the capacitance associated with the surface depletion region at its maximum width. (A more detailed description of the behavior of the high- and low-frequency capacitances under inversion conditions will be given in a later section.) Figures 3.5 and 3.6 show typical capacitance versus voltage relationships for MOS capacitors fabricated on *p*- and *n*-type silicon substrates, respectively. Both the high- and the low-frequency curves are shown.

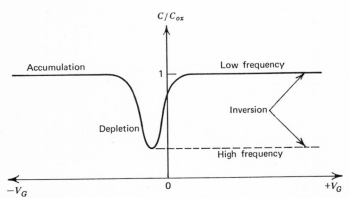

FIGURE 3.5 Typical capacitance-voltage relationship for a MOS capacitor fabricated on a *p*-type silicon substrate.

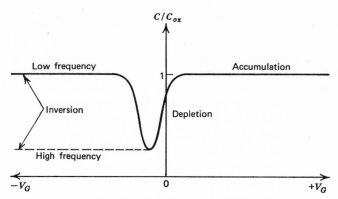

FIGURE 3.6 Typical capacitance-voltage relationship for a MOS capacitor fabricated on an n-type silicon substrate.

3.2   THE THEORY OF THE MOS CAPACITOR

It was seen in the previous section that the modulation of the surface space charge region by the gate field results in a variation of the gate-to-substrate capacitance with gate voltage. A simple physical model for the dependence of this capacitance on the applied gate voltage was first proposed by Grove, Deal, Snow, and Sah.[1] Extremely good correlation has been observed between the model and experimental high- and low-frequency capacitance-voltage curves. The following sections are based on the work above.

3.2.1   The Effects of Fixed Positive Interface Charge Density and Metal-Semiconductor Work Function Difference on the Characteristics of MOS Capacitors

Both the presence of the fixed positive charge density located at the oxide-silicon interface and the work function difference between the gate electrode and the silicon substrate will have a pronounced effect on the surface space charge region of a MOS device. As previously mentioned, the existence of a positive interface charge density per unit area, Q_{SS}, has been experimentally observed on both n- and p-type silicon wafers over a wide range of substrate resistivities and crystalline orientations. (The actual physical nature of Q_{SS} is discussed in a later chapter.) Since this charge density is always positive, it will always induce an equal and opposite "mirror" charge density at the silicon surface. Thus with increasing values of Q_{SS}, the onset of inversion for MOS capacitor structures fabricated on *either p-* or n-type silicon substrates will always be observed at increasingly *negative* gate voltages.

The characteristics of the surface space charge region will also be influenced by the dissimilarity of the work functions of the (metal) gate electrode and the silicon substrate. The effect of this dissimilarity will be to produce a certain amount of band-bending at the silicon surface, even with zero gate voltage. As discussed in Section 2.2.1, the amount of gate voltage required to compensate for the differences in the work functions and, in the absence of any interface charge density, bring the silicon surface to a "flat-band" condition is denoted by $\phi_{MS'}$. When the effect of Q_{SS} is also taken into account, the *flatband voltage* (which is defined as the amount of gate-to-substrate voltage which is required *to produce zero band-bending at the silicon surface*) is given simply by

$$V_{FB} = \phi_{MS'} - \frac{Q_{SS}}{C_{ox}}. \tag{3.3}$$

3.2.2   The Dependence of Gate-to-Substrate Capacitance on the Gate Voltage

The applied gate voltage is related to the surface potential ϕ_s, the voltage impressed across the insulator V_{OX}, and the metal-semiconductor work function difference $\phi_{MS'}$ by

$$V_G = V_{OX} + \phi_{MS'} + \phi_s. \tag{2.42}$$

Once again, since all the electric field resulting from the application of the gate-to-substrate voltage will lie totally within the region consisting of the gate electrode, the gate insulator, and the surface of the semiconductor, the charge neutrality condition in this region is

$$Q_G + Q_{SS} + Q_{\text{total}} = 0, \tag{3.4}$$

where Q_{total} is given by

$$Q_{\text{total}} \equiv Q_{\text{inv}} + Q_{SD}. \tag{2.18}$$

The charge density per unit area on the gate electrode will be related to V_{OX} by

$$Q_G = C_{ox} V_{OX}. \tag{2.43}$$

Combining (2.42), (3.4), and (2.43) yields

$$V_G - \phi_{MS'} + \frac{Q_{SS}}{C_{ox}} = \phi_s - \frac{Q_{\text{total}}}{C_{ox}}. \tag{3.5}$$

Equation 3.5 can be expressed in terms of the flatband voltage as

$$V_G - V_{FB} = \phi_s - \frac{Q_{\text{total}}}{C_{ox}}. \tag{3.6}$$

The gate-to-substrate capacitance per unit area can be written as

$$C = \frac{dQ_G}{dV_G}, \tag{3.7}$$

and the differential of the gate voltage can be obtained directly from (2.42). Thus

$$C = \frac{dQ_G}{dV_{OX} + d\phi_s} = \left(\frac{dV_{OX}}{dQ_G} + \frac{d\phi_s}{dQ_G}\right)^{-1}. \tag{3.8}$$

dQ_G/dV_{OX} is simply the oxide capacitance per unit area, C_{ox}. From (3.4), Q_G can be expressed as

$$Q_G = -Q_{\text{total}} - Q_{SS}. \tag{3.9}$$

Therefore,

$$\frac{dQ_G}{d\phi_s} = -\frac{dQ_{\text{total}}}{d\phi_s} - \frac{dQ_{SS}}{d\phi_s}. \tag{3.10}$$

The capacitance per unit area associated with the surface space charge region of the semiconductor can be defined as

$$C_s = -\frac{dQ_{\text{total}}}{d\phi_s}, \tag{3.11}$$

and, similarly, the capacitance per unit area associated with the interface charge density can be defined as

$$C_{SS} = -\frac{dQ_{SS}}{d\phi_s}. \tag{3.12}$$

Combining (3.8), (3.10), (3.11), and (3.12) yields an expression for the gate-to-substrate capacitance per unit area in terms of C_{ox}, C_s, and C_{SS}. The result is

$$C = \left[\left(\frac{1}{C_{ox}}\right) + \left(\frac{1}{C_s + C_{SS}}\right)\right]^{-1}. \tag{3.13}$$

The fixed positive interface charge density is independent of the surface potential and, if it is assumed that no voltage-dependent trapping mechanisms are occurring at the oxide-silicon interface, then C_{SS} will go to zero and the gate-to-substrate capacitance per unit area will be given by

$$C = \left[\left(\frac{1}{C_{ox}}\right) + \left(\frac{1}{C_s}\right)\right]^{-1}. \tag{3.14}$$

Thus the equivalent circuit of a MOS capacitor is as shown in Figure 3.7. The C_s will be a function of frequency, and its low- and high-frequency behavior is discussed in greater detail in the following sections.

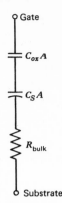

FIGURE 3.7   Simple equivalent circuit for a MOS capacitor.

3.2.3   Variation of Gate-to-Substrate Capacitance as a Function of Frequency

Under accumulation or depletion conditions, the charge required to terminate the gate field at the semiconductor surface will be supplied by the movement of majority carriers from the substrate. Consequently, the charge induced at the surface of the semiconductor will follow a high-frequency a.c. signal applied to the gate electrode as long as

$$f \ll \frac{1}{\tau_D},\tag{3.15}$$

where τ_D is the dielectric relaxation time associated with the semiconductor substrate. (The dielectric relaxation time of a material is given by the product of its dielectric constant times its resistivity. For 5-Ω-cm silicon, for example, τ_D is equal to 5×10^{-12} sec.)

Under inversion conditions, the situation is entirely different, and the *minority carriers* in the substrate must now provide the charge required to terminate the gate field. Hofstein and Warfield[2] have shown that the minority carriers needed to charge the inversion layer can be obtained from any one of the following mechanisms acting alone or in combination with one another:

1. Generation-recombination within the depletion region directly below the inversion layer.

2. Generation-recombination through surface states located at the insulator-semiconductor interface.

3. Diffusion of minority carriers to the edge of the depletion region, which drift across this region into the inversion layer.

A separate current can be associated with each of the mechanisms above that will contribute to the charging of the surface inversion layer. However, since the magnitude of each current will be relatively small, a considerable amount of time will be required to charge the inversion layer even if all three mechanisms are active. Hence the response time of the inversion layer will be much longer than the dielectric relaxation time. (Typical inversion layer response times are on the order of tenths or hundredths of a second.)

As was pointed out in Section 3.1.3, the capacitance of the MOS structure under inversion conditions will be a strong function of whether the charge in the inversion layer can follow the a.c. variation in the applied gate voltage. If the response time of the surface inversion layer is denoted by τ_{inv}, then the inversion layer can follow a signal of frequency f applied at the gate electrode if

$$f \ll \frac{1}{\tau_{inv}}. \qquad (3.16)$$

However, for frequencies such that

$$f \gg \frac{1}{\tau_{inv}}, \qquad (3.17)$$

the inversion layer will not follow the gate signal and a nonequilibrium situation will exist. Consequently, *while the capacitance-versus-voltage curve associated with a MOS capacitor will be independent of frequency under accumulation and depletion conditions for all frequencies of practical interest, this will not be true when a surface inversion layer is present.* The low- and high-frequency behavior of the MOS capacitance for the limiting cases described by (3.16) and (3.17), respectively, is discussed in the next two sections.

In most MOS devices, the dominant mechanism that determines the inversion layer response time is usually that of generation-recombination at centers located within the depletion region. Since an increase in the operating temperature of the device or the application of incident light will increase the carrier generation-recombination rate, thereby increasing the current associated with this mechanism, a corresponding decrease in the inversion layer response time will result. Consequently, under these conditions, the requirement for thermal equilibrium given by (3.16) will be satisfied for higher frequencies and this will in turn be reflected in the observed capacitance-voltage characteristics of the device.

3.2.4 Gate-to-Substrate Capacitance at Low Frequencies

For frequencies such that $f \ll \tau_{inv}^{-1}$, the charge in the inversion layer can follow the gate signal and a condition of thermal equilibrium will exist.

Remembering that the total surface space charge density per unit area for a MOS device fabricated on a p-type substrate is given by (2.19), it follows that the capacitance per unit area associated with the surface space charge region can be obtained through (2.19) and (3.11). Simple differentiation yields

$$C_s = -\frac{dQ_{\text{total}}}{d\phi_s}$$

$$\cong \frac{q^2 L_D(n_s - p_s + N_A - N_D)}{kT\{2[\cosh(u_s - u_F) - \cosh(u_F) + u_s \sinh(u_F)]\}^{1/2}} \tag{3.18}$$

where

$$n_s \cong n_i e^{(u_s - u_F)}$$

and

$$p_s \cong n_i e^{-(u_s - u_F)} \tag{3.19}$$

are the carrier concentrations at the surface of the semiconductor ($x = 0$) and

$$N_A \cong n_i e^{u_F}$$
$$N_D \cong n_i e^{-u_F}. \tag{3.20}$$

Equation 3.18 can be expressed in terms of Q_{total} as

$$C_s = \frac{\epsilon_s q(p_s - n_s + N_D - N_A)}{Q_{\text{total}}}. \tag{3.21}$$

The low-frequency (equilibrium) capacitance-voltage characteristics of a MOS capacitor may now be calculated through (2.19), (3.6), (2.10), (2.12), (3.14), and (3.21), as done in the following manner:

1. u_F is determined from the doping concentration in the semiconductor substrate in (3.20).
2. Q_{total} is obtained from (2.19) for a specific value of u_s.
3. $V_G - V_{FB}$ is now calculated from (3.6) for the specific value of u_s.
4. The space charge capacitance per unit area, C_s, is obtained by evaluating (3.21).
5. The gate-to-substrate capacitance per unit area, C, is calculated using (3.14) for the specific value of u_s.
6. The procedure above is repeated for a number of different values of u_s over the range of interest. For each value, the calculated magnitude of C is plotted versus $V_G - V_{FB}$.

Theoretical MOS capacitance-voltage relationships have been prepared in the manner above, with the aid of a computer, by Goetzberger.[3] The shape of these curves will vary as a function of both substrate doping concentration and gate oxide thickness. Typical theoretical capacitance-voltage curves for

MOS capacitors fabricated on p-type silicon substrates with acceptor doping concentrations ranging from $1 \times 10^{14}/\text{cm}^3$ to $1 \times 10^{17}/\text{cm}^3$ as a function of the thickness of the silicon dioxide gate insulator are shown in Figures 3.8 through 3.11. The values of gate-to-substrate capacitance obtained have been normalized to the oxide capacitance per unit area and the normalized capacitance is plotted versus the gate voltage minus the flatband voltage. The solid curves correspond to the theoretical low-frequency capacitance-voltage relationships. The dashed curves indicate the high-frequency behavior of the normalized capacitance for values of gate voltage minus the flatband voltage sufficient to cause surface inversion. (The behavior of the high-frequency curves is discussed in the next section.) Figures 3.12 through 3.15 illustrate the behavior of the surface potential $(u_s \equiv \psi_s)$ as a function of the gate voltage minus the flatband voltage, again for different values of both silicon dioxide gate insulator thickness and acceptor doping concentration in the p-type silicon substrate. (The data shown in these figures were obtained directly from step 3 of the procedure used to calculate the behavior of the gate-to-substrate capacitance, see p. 58.) Although the curves of Figures 3.8 through 3.15 are for MOS capacitors fabricated on p-type silicon substrates, they are easily converted for n-type silicon substrates merely by changing the sign of the voltage axis.

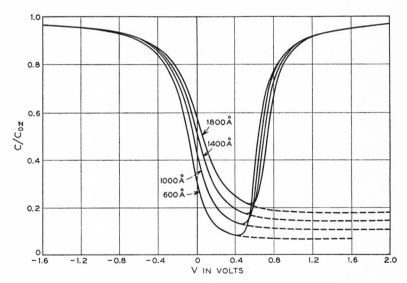

FIGURE 3.8 Theoretical capacitance-voltage relationships for MOS capacitors fabricated on p-type silicon with $N_A = 1 \times 10^{14}/\text{cm}^3$, as a function of different thicknesses of silicon dioxide gate insulator. (After Goetzberger[3]; reprinted with the permission of the American Telephone and Telegraph Company.)

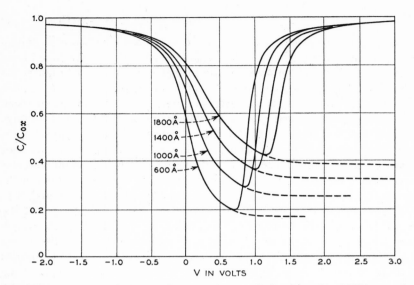

FIGURE 3.9 Theoretical capacitance-voltage relationships for MOS capacitors fabricated on p-type silicon with $N_A = 1 \times 10^{15}/\text{cm}^3$, as a function of different thicknesses of silicon dioxide gate insulator. (After Goetzberger[3]; reprinted with the permission of the American Telephone and Telegraph Company.)

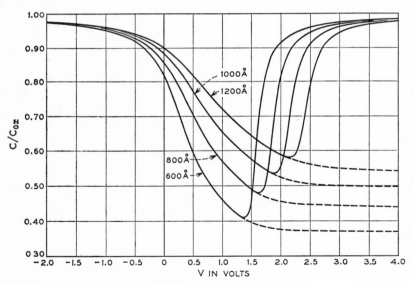

FIGURE 3.10 Theoretical capacitance-voltage relationships for MOS capacitors fabricated on p-type silicon with $N_A = 1 \times 10^{16}/\text{cm}^3$, as a function of different thicknesses of silicon dioxide gate insulator. (After Goertzberger[3]; reprinted with the permission of the American Telephone and Telegraph Company.)

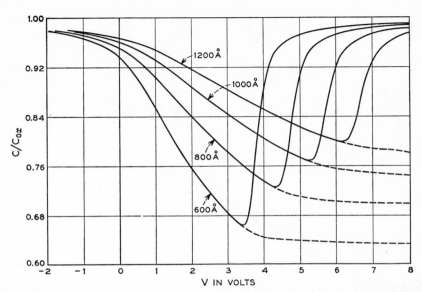

FIGURE 3.11 Theoretical capacitance-voltage relationships for MOS capacitors fabricated on p-type silicon with $N_A = 1 \times 10^{17}/cm^3$, as a function of different thicknesses of silicon dioxide gate insulator. (After Goetzberger[3]; reprinted with the permission of the American Telephone and Telegraph Company.)

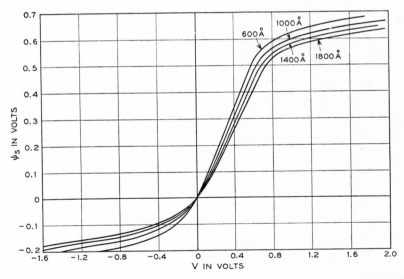

FIGURE 3.12 Theoretical values of surface potential for MOS capacitors fabricated on p-type silicon with $N_A = 1 \times 10^{14}/cm^3$, as a function of different thicknesses of silicon dioxide. (After Goetzberger[3]; reprinted with the permission of the American Telephone and Telegraph Company.)

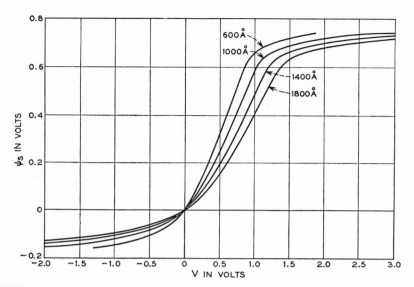

FIGURE 3.13 Theoretical values of surface potential for MOS capacitors fabricated on p-type silicon with $N_A = 1 \times 10^{15}/cm^3$, as a function of different thicknesses of silicon dioxide. (After Goetzberger[3]; reprinted with the permission of the American Telephone and Telegraph Company.)

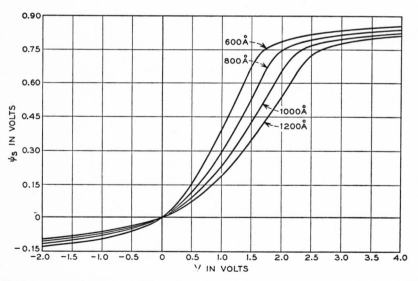

FIGURE 3.14 Theoretical values of surface potential for MOS capacitors fabricated on p-type silicon with $N_A = 1 \times 10^{16}/cm^3$, as a function of different thicknesses of silicon dioxide. (After Goetzberger[3]; reprinted with the permission of the American Telephone and Telegraph Company.)

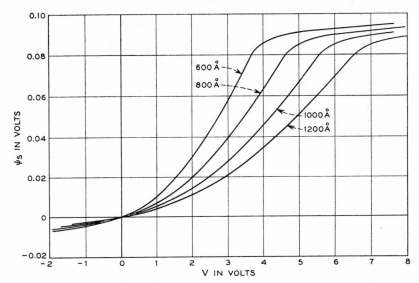

FIGURE 3.15 Theoretical values of surface potential for MOS capacitors fabricated on p-type silicon with $N_A = 1 \times 10^{17}/cm^3$, as a function of different thicknesses of silicon dioxide. (After Goetzberger[3]; reprinted with the permission of
the American Telephone and Telegraph Company.)

It can easily be seen in Figures 3.8 through 3.11 that the theoretical low-
frequency capacitance-voltage curves are consistent with the qualitative
explanation of their behavior as described in Section 3.1. Under accumulation conditions, the capacitance per unit area is approximately equal to the
dielectric capacitance per unit area, C_{ox}. When the surface depletion region
forms, the space charge capacitance adds in series to the dielectric capacitance; consequently, the total gate-to-substrate capacitance decreases. If the
d.c. bias on the gate electrode becomes sufficiently positive such that an
inversion layer forms at the surface of the silicon, the charge in the inversion
layer can follow a low-frequency gate signal and the charge in the surface
depletion region will remain constant at its maximum value. Hence for
low-frequency gate signals,

$$\frac{dQ_{SD}}{dV_G} \to 0 \tag{3.22}$$

under inversion conditions. As a result, any *small-signal* variations in the
gate field will now be accommodated by fluctuations in the charge stored in the
surface inversion layer and the gate-to-substrate capacitance will rise again
and approach C_{ox}.

3.2.5   Gate-to-Substrate Capacitance at High Frequencies

For frequencies such that $f \gg \tau_{\text{inv}}^{-1}$, the charge in the surface inversion layer will not follow the gate signal. Thus for high-frequency gate signals, the results of Section 3.2.4 (which were based upon the assumption that the system was in thermal equilibrium) will no longer be valid. Since the charge density in the inversion layer cannot follow the high-frequency variations in the gate voltage, Q_{inv} can be assumed to be constant for a given d.c. bias. In other words,

$$\frac{dQ_{\text{inv}}}{dV_G} \to 0 \tag{3.23}$$

under inversion conditions and the a.c. field will have to be terminated with charge associated with the *surface depletion region*. Under these conditions, the depletion region charge density per unit area and the width of the depletion region will fluctuate around their quiescent values $Q_{SD_{\max}}$ and $x_{d_{\max}}$, respectively, following the high-frequency variations in the gate voltage such that the condition for charge neutrality, as given by (2.18) and (3.4), is satisfied. The high-frequency gate-to-substrate capacitance per unit area will be equal to the oxide capacitance per unit area, C_{ox}, in series with the (minimum) capacitance per unit area associated with the maximum width of the depletion region. It can easily be seen from (3.23) that the inversion layer will not contribute to the high-frequency capacitance.

The capacitance of the surface space charge region when the depletion region is at its maximum width will be denoted by $C_{SD_{\min}}$ and will be given by

$$C_{SD_{\min}} = -\frac{dQ_{SD_{\max}}}{d\phi_s}. \tag{3.24}$$

Substitution of (2.30) and (2.31) yields

$$C_{SD_{\min}} = qN_A \frac{dx_{d_{\max}}}{d\phi_s} = qN_A \left(\frac{\epsilon_s}{2qN_A\phi_s}\right)^{1/2}, \tag{3.25}$$

or

$$C_{SD_{\min}} = \frac{\epsilon_s}{x_{d_{\max}}}. \tag{3.26}$$

Physically, $C_{SD_{\min}}$ can be thought of as consisting of a parallel-plate capacitor of unit area with dielectric constant ϵ_s and a plate separation equal to the maximum width of the surface depletion region. Consequently, for *high-frequency gate signals*, the gate-to-substrate capacitance per unit area under inversion conditions will *not* rise and approach C_{ox} as in the low-frequency case. Instead, it will remain constant at its minimum value, independent of d.c. gate bias and will be given by

$$C = \left[\left(\frac{1}{C_{ox}}\right) + \left(\frac{1}{C_{SD_{min}}}\right)\right]^{-1}. \qquad (3.27)$$

High-frequency (nonequilibrium) capacitance-voltage relationships for MOS capacitors fabricated on p-type silicon substrates can now be calculated by using (3.25) to describe the inversion capacitance per unit area of the surface space charge region in the high-frequency limit [instead of (3.21)]. The procedure is very similar to the approach outlined in the previous section. Theoretical high-frequency capacitance-voltage relationships calculated in this manner are also shown in Figures 3.8 through 3.11. The dashed lines correspond to the high-frequency behavior under inversion conditions. Similar high-frequency curves for devices fabricated on n-type substrates can be obtained by an analogous procedure.

Although the inversion layer of a MOS capacitor cannot follow the high-frequency component of the gate voltage, it should be noted that this is not the case for a MOS field-effect transistor. Under normal operating conditions, the heavily doped source region of the MOSFET will always be in direct ohmic contact with the inversion layer. Thus the source region can easily supply the inversion layer with the charge required to follow the high-frequency gate signal. Consequently, unlike the MOS capacitor structure in which the response time of the inversion layer is limited by the amount of charge that can be supplied by the three mechanisms discussed in Section 3.2.3, the MOSFET is theoretically capable of operating at much higher frequencies.

3.2.6 Gate-to-Substrate Capacitance Under Deep-Depletion Conditions

If the gate insulator is somewhat "leaky" and if even a small amount of steady-state current can flow from gate to substrate, or if the gate voltage is switched very rapidly from the bias range corresponding to accumulation to a value which would normally result in surface inversion under steady-state conditions, the MOS capacitor structure will operate in the *deep-depletion* mode. If, as in the former case, minority carriers are unable to accumulate at the surface of the silicon, or if the capacitance is measured before minority carriers have a chance to accumulate at the silicon surface, as in the latter (transient) situation, then the gate field will have to be terminated instead with charge associated with the *surface depletion region*. Under these conditions, instead of surface inversion taking place, the width of the surface depletion region will continue to increase beyond its maximum equilibrium value, $x_{d_{max}}$. The measured gate-to-substrate capacitance will, consequently, decrease with increasing gate field in the bias range normally corresponding to inversion. This is illustrated in Figure 3.16, where the behavior of the gate-to-substrate capacitance of a MOS capacitor fabricated on a p-type silicon substrate is shown under deep-depletion conditions, as well as for both the high- and low-frequency cases when an n-type surface inversion layer is present.

The gate-to-substrate capacitance per unit area under depletion conditions is given by (3.1), which can be rewritten as

$$\frac{C}{C_{ox}} = \left[1 + \left(\frac{C_{ox}}{C_{SD}}\right)\right]^{-1},\qquad(3.28)$$

where the capacitance per unit area associated with the surface depletion region, C_{SD}, is defined by (3.2). When only a surface depletion region is present, $Q_{total} = Q_{SD}$ and (3.6) takes the form

$$\phi_s = (V_G - V_{FB}) + \frac{Q_{SD}}{C_{ox}}.\qquad(3.29)$$

Poisson's equation can be integrated twice, from $x = 0$ to $x = x_d$, with $\rho(x) = -qN_A$. The result is

$$\phi_s = \int_0^{x_d} \frac{qN_A x \, dx}{\epsilon_s} = \frac{+qN_A x_d^2}{2\epsilon_s} = \frac{+qN_A \epsilon_s}{2C_{SD}^2}.\qquad(3.30)$$

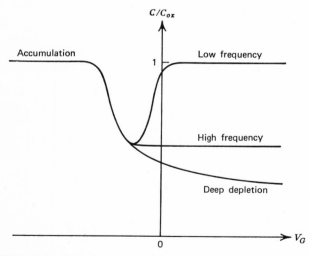

FIGURE 3.16 Behavior of a MOS capacitor fabricated on p-type silicon in the bias range corresponding to surface inversion: low-frequency capacitance, high-frequency capacitance, and deep-depletion capacitance.

Remembering that the charge density per unit area contained in the surface depletion region is given by (2.33), then (3.29) reduces to

$$\phi_s = (V_G - V_{FB}) - \frac{qN_A x_d}{C_{ox}}, \tag{3.31}$$

or

$$\phi_s = (V_G - V_{FB}) - \frac{qN_A \epsilon_s}{C_{SD}C_{ox}}. \tag{3.32}$$

Equating (3.32) and (3.30) yields

$$\left(\frac{1}{C_{SD}C_{ox}} + \frac{1}{2C_{SD}^2}\right) - \frac{(V_G - V_{FB})}{qN_A \epsilon_s} = 0, \tag{3.33}$$

and after multiplying by $2C_{ox}^2$ and rearranging, the result is

$$\left(\frac{C_{ox}}{C_{SD}}\right)^2 + \left(\frac{2C_{ox}}{C_{SD}}\right) - \frac{2C_{ox}^2(V_G - V_{FB})}{qN_A \epsilon_s} = 0. \tag{3.34}$$

Now, (3.28) can be written as

$$\left(\frac{C_{ox}}{C}\right) = \left(\frac{C_{ox}}{C_{SD}}\right) + 1, \tag{3.35}$$

or

$$\left(\frac{C_{ox}}{C_{SD}}\right) = \left(\frac{C_{ox}}{C}\right) - 1 \equiv \gamma. \tag{3.36}$$

Substituting (3.36) into (3.34), one obtains

$$\gamma^2 + 2\gamma - \frac{2C_{ox}^2(V_G - V_{FB})}{qN_A \epsilon_s} = 0. \tag{3.37}$$

By definition, γ must be positive. Therefore,

$$\gamma = -1 + \left[1 - \frac{2C_{ox}^2(V_G - V_{FB})}{qN_A \epsilon_s}\right]^{1/2} = \frac{C_{ox}}{C_{SD}}. \tag{3.38}$$

Finally, substituting (3.38) into (3.28) yields an expression for the (normalized) gate-to-substrate capacitance of a MOS capacitor in deep-depletion:

$$\frac{C}{C_{ox}} = \left\{\left[1 - \frac{2C_{ox}^2(V_G - V_{FB})}{qN_A \epsilon_s}\right]^{1/2}\right\}^{-1}. \tag{3.39}$$

3.3 CORRELATION BETWEEN THEORETICAL AND EXPERIMENTAL MOS CAPACITANCE-VOLTAGE CHARACTERISTICS

The study of the characteristics of capacitance-versus-voltage relationships for MOS capacitor structures has become one of the most powerful tools available to researchers in the study of semiconductor surfaces. The use of this technique routinely requires only the fabrication of the relatively simple MOS capacitor structure, while virtually all of the results obtained are directly applicable to the more complex MOSFET. The basis of the technique is the fact that the important electrical properties of a semiconductor-insulator interface can be determined simply by comparing the theoretical capacitance-voltage relationship of a MOS capacitor fabricated on the semiconductor substrate with the actual experimentally observed C-V curve.

Figure 3.17 illustrates a comparison of the theoretical and typical experimental high-frequency capacitance-versus-voltage curves for a MOS capacitor fabricated on a p-type silicon substrate. The theoretical curve, marked a, is obtained by following the procedure which was described in Section 3.2.5 and setting the flatband voltage, V_{FB}, equal to zero. Curve c is obtained experimentally by measuring the gate-to-substrate capacitance at a high frequency while slowly varying the d.c. bias and tracing out the results on an x-y recorder. The experimental curve is shifted in the negative direction

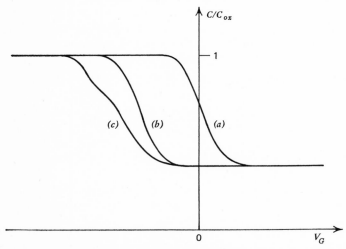

FIGURE 3.17 Comparison of theoretical and experimental capacitance versus voltage curves for a MOS capacitor fabricated on p-type silicon: (a) denotes theoretical curve, (b) denotes experimental curve in the absence of fast surface states, and (c) denotes experimental curve with fast surface states present.

with respect to the theoretical curve, which is almost always the case and is a result of the fixed positive interface charge density always being a positive quantity. Curve b illustrates an experimental curve which exhibits a *parallel* shift from the theoretical curve. This parallel shift will only occur in the absence of fast surface states at the oxide-silicon interface and can be attributed to the effects of Q_{SS}, any ionic charge present in the gate insulator, and the work function term, $\phi_{MS'}$. The voltage shift between curves a and b will be equal to

$$\Delta V_{G_{ab}} = -\frac{Q_{SS}}{C_{ox}} - \frac{\int_{x=-T_{ox}}^{x=0-} [1 + (x/T_{ox})]\rho(x)\, dx}{C_{ox}} + \phi_{MS'}. \quad (3.40)$$

Occasionally, the actual slope of the experimental C-V curve for gate voltages corresponding to depletion conditions may be slightly less than that of the theoretical curve. This can be seen, for example, in the experimental curve marked c. The decrease in the slope of the experimental curve can be directly attributed to the charging and discharging of any fast surface states that may be present at the oxide-silicon interface. Essentially, the fast surface states are surface recombination-generation centers (with energies lying within the forbidden region between the conductance and valence bands of the semiconductor) that result from the lattice disorder that occurs near the interface. The fast surface state charge density per unit area differs from Q_{SS} in that the charges present in the fast states are easily exchanged with the bulk of the semiconductor. Specifically, the total charge density in fast surface states will be a function of the surface potential and, consequently, will vary with the gate voltage. If the density of *charged* fast states per unit area is denoted by N_{ST} then, following the procedure which was used in deriving (3.13), the gate-to-substrate capacitance per unit area can be shown to be

$$C = \left[\left(\frac{1}{C_{ox}}\right) + \left(\frac{1}{C_s + C_{ST}}\right)\right]^{-1}, \quad (3.41)$$

where

$$C_{ST} \equiv -q\,\frac{dN_{ST}}{d\phi_s}. \quad (3.42)$$

It follows that the amount of *additional* shift of the experimental C-V curve with respect to the theoretical curve that is attributable to fast surface states will also vary with the applied gate voltage.

With increasingly negative gate voltage, a growing number of electrons will be repelled out of the fast surface states and the states, which were originally electrically neutral, will become positively charged. It can easily be seen from (3.41) and (3.42) that once the states are positively charged they can contribute to C_{ST}, and will affect the total gate-to-substrate capacitance.

As a result, the observed experimental C-V curve shifts further to the left with increasingly negative gate voltage, as shown in Figure 3.17. The amount of shift as a function of gate voltage that is a result of the presence of fast surface states will be

$$\Delta V_{G_{bc}} = - \frac{q N_{ST}(V_G)}{C_{ox}},$$ (3.43)

and the total shift between the experimental curve c and the theoretical curve will be equal to

$$\Delta V_{G_{ac}} = \Delta V_{G_{ab}} - \frac{q N_{ST}(V_G)}{C_{ox}}.$$ (3.44)

While the presence of fast surface states had a pronounced effect on the electrical characteristics of early MOS devices, it should be noted that recent advances in silicon technology have succeeded in lowering the density of fast surface states to the point where their effect on the observed device characteristics is usually negligible.

REFERENCES

1. A. S. Grove, B. E. Deal, E. H. Snow, and C. T. Sah, Investigation of Thermally Oxidized Silicon Surfaces Using Metal-Oxide-Semiconductor Structures, *Solid State Electronics*, Vol. 8, 1965, pp. 145–163.

2. S. R. Hofstein and G. Warfield, Physical Limitations on the Frequency Response of a Semiconductor Surface Inversion Layer, *Solid State Electronics*, Vol. 8, 1965, pp. 321–341.

3. A. Goetzberger, Ideal MOS Curves for Silicon, *Bell System Technical Journal*, Vol. 45, 1966, pp. 1097–1122.

BIBLIOGRAPHY

Brotherton, S. D., and P. Burton, The Influence of Non-Uniformly Doped Substrates on MOS C-V Curves, *Solid State Electronics*, Vol. 13, 1970, pp. 1591–1595.

Bulucea, C. D., Investigation of Deep-Depletion Regime of MOS Structures Using Ramp Response Method, *Electronics Letters*, Vol. 6, No. 15, 1970, pp. 479–481.

Goetzberger, A., Ion Drift in the Fringing Field of MOS Capacitors, *Solid State Electronics*, Vol. 9, 1966, pp. 871–878.

Goetzberger, A., and J. C. Irvin, Low-Temperature Hysteresis Effects in Metal-Oxide-Silicon Capacitors Caused by Surface-State Trapping, *IEEE Transactions on Electron Devices*, Vol. ED-15, No. 12, 1968, pp. 1009–1014.

Goetzberger, A., and E. Klausmann, A Capacitance Mapping Technique for Investigation of Localized Recombination-Generation Sites in Si-SiO$_2$ Interfaces, *Proceedings of the IEEE*, Vol. 58, No. 5, May 1970, pp. 799–800.

Goetzberger, A., and E. H. Nicollian, MOS Avalanche and Tunneling Effects in Silicon Surfaces, *Journal of Applied Physics*, Vol. 38, No. 12, 1967, pp. 4582–4588.

Grove, A. S., E. H. Snow, B. E. Deal, and C. T. Sah, Simple Physical Model for the Space-Charge Capacitance of Metal-Oxide-Semiconductor Structures, *Journal of Applied Physics*, Vol. 35, No. 8, August 1964, pp. 2458–2460.

Heiman, F. P., and G. Warfield, The Effects of Oxide Traps on the MOS Capacitance, *IEEE Transactions on Electron Devices*, Vol. ED-12, No. 4, April 1965, pp. 167–178.

Hielscher, F. H., and H. M. Preier, Non-Equilibrium C-V and I-V Characteristics of Metal-Insulator-Semiconductor Capacitors, *Solid State Electronics*, Vol. 12, 1969, pp. 527–538.

Koomen, J., The Measurement of Interface State Charge in the MOS System, *Solid State Electronics*, Vol. 14, 1971, pp. 571–580.

Kuhn, M., A Quasi-Static Technique for MOS C-V and Surface State Measurements, *Solid State Electronics*, Vol. 13, 1970, pp. 873–885.

Lehovec, K., Rapid Evaluation of C-V plots for MOS Structures, *Solid State Electronics*, Vol. 11, 1968, pp. 135–137.

Lindner, E. R., Limitations of the MIS Capacitance Method Resulting from Semiconductor Properties, *Solid State Electronics*, Vol. 13, 1970, pp. 1597–1605.

Nicollian, E. H.. and A. Goetzberger, Lateral AC Current Flow Model for Metal-Insulator-Semiconductor Capacitors, *IEEE Transactions on Electron Devices*, Vol ED-12, No. 3, March 1965, pp. 108–117.

Nicollian, E. H., A. Goetzberger, and C. N. Berglund, Avalanche Injection Currents and Charging Phenomena in Thermal SiO_2, *Applied Physics Letters*, Vol. 15, No. 6, 1969, pp. 174–177.

Pierret, R. F., A. Cabana, and C. T. Ho, Large ac Signal Behavior of the MOS Capacitor Biased in Inversion, *IEEE Transactions on Electron Devices*, Vol. ED-17, No. 7, July 1970, pp. 561–562.

Poirier, R., and J. Olivier, Hot Electron Emission from Silicon into Silicon Dioxide by Surface Avalanche, *Applied Physics Letters*, Vol. 15, No. 11, 1969, pp. 364–365.

Prince, J. L., J. J. Wortman, and J. R. Hauser, On the Deeply Depleted MOS Capacitor, *Proceedings of the IEEE*, Vol. 58, No. 5., May 1970, pp. 842–844.

Whelan, M. V., Graphical Relations Between Surface Parameters of Silicon, to be Used in Connection with MOS Capacitance Measurements, *Philips Research Reports*, Vol. 20, 1965, pp. 620–632.

Whelan, M. V., Influence of Charge Interactions on Capacitance Versus Voltage Curves in MOS Structures, *Philips Research Reports*, Vol. 20, 1965, pp. 562–577.

Zaininger, K. H., Automatic Display of C-V Curves for Metal-Insulator-Semiconductor (MIS) Structures, *Proceedings of the IEEE*, Vol. 54, No. 7, July, 1966, pp. 1001–1002.

Zaininger, K. H., and F. P. Heiman, The C-V Technique as an Analytical Tool, *Solid State Technology*, May 1970 (Part 1), pp. 49–56.

Zaininger, K. H., and F. P. Heiman, The C-V Technique as an Analytical Tool, *Solid State Technology*, June 1970 (Part 2), pp. 46–55.

Zaininger, K. H., and G. Warfield, Limitations of the MOS Capacitance Method for the Determination of Semiconductor Surface Properties, *IEEE Transactions on Electron Devices*, Vol. ED-12, No. 4, April, 1965, pp. 179–193.

PROBLEMS

3.1 Discuss the behavior of the capacitance versus voltage relationship of a MOS capacitor as a function of temperature. In particular, discuss the variation with temperature of the ratio of the minimum capacitance at the onset of strong inversion to the oxide capacitance (C_{min}/C_{ox}) and discuss the effect of temperature on the inversion capacitance over the entire frequency range.

3.2 Discuss the difference between the flatband voltage and the threshold voltage. Why is the threshold voltage dependent on the charge density in the surface depletion region while the flatband voltage is not?

3.3 Derive an expression for the minimum high-frequency capacitance per unit area associated with a MOS capacitor structure as a function of the thickness of the silicon dioxide gate insulator and the effective doping concentration in the silicon substrate.

3.4 An aluminum-gate MOS capacitor is fabricated on a p-type silicon substrate with an effective acceptor doping concentration of $N_A = 10^{15}/cm^3$. The thickness of the silicon dioxide gate insulator is 1000 Å and the fixed positive interface charge density per unit area divided by the electronic charge is approximately $1.5 \times 10^{11}/cm^2$. Assuming that the amount of ionic charge density within the gate insulator is negligible, (a) plot the behavior of the high-frequency C-V curve at room temperature if no fast surface states are present at the silicon-silicon dioxide interface. (b) Plot the behavior of the high-frequency C-V curve at room temperature if N fast surface states per centimeter^2, having a single energy $E = \frac{1}{2}E_G$, are present at the silicon-silicon dioxide interface. Assume these fast states have the characteristics described in Section 3.3, that is, they will be electrically neutral when occupied by an electron but will be positively charged when vacant. (c) Repeat part b assuming that the states will be electrically neutral when vacant but will become negatively charged when occupied by an electron. (d) Repeat parts b and c for fast states with a uniform distribution in energy of M states/(cm^2)/(eV) over the forbidden band gap.

3.5 Discuss the effect upon the high-frequency C-V curve of diffusing an n^+ guard-band into the silicon surface around the periphery of a MOS capacitor fabricated on a p-type substrate.

3.6 Keeping the n^+ guard-band described in the previous problem at ground potential, a negative bias is now applied to the substrate. What is the effect on both the high- and the low-frequency C-V curves? Explain the reason for this effect. Qualitatively plot the behavior of the low-frequency C-V curve for increasingly negative substrate voltages. Derive an expression for the minimum capacitance of the structure as a function of the substrate-to-guard band voltage.

4

Three-Terminal
Characteristics of MOS
Field-Effect Transistors

4.1 THE BEHAVIOR OF THE DRAIN CURRENT WITH INCREASING DRAIN-TO-SOURCE VOLTAGE

As discussed in Chapter 1, the three-terminal electrical characteristics associated with typical MOSFETs can, in general, be divided into three distinct operating regions. For small values of applied drain-to-source potential, the drain current is found to increase linearly with drain voltage and the device is said to be operating in the *variable resistance region*. As the drain voltage is increased, the drain current gradually saturates to a relatively constant level, at a given gate voltage, and is then observed to increase only slightly with increasing drain voltage. Under these conditions, the MOSFET is considered to be operating in the *region of saturated current flow*. Finally, if the applied drain voltage is increased to a sufficiently high value, the drain current is found to increase rapidly as the drain junction begins to avalanche, and the device will enter the *avalanche breakdown region*. This chapter treats the behavior of MOSFETs in each of the regions of operation above. In addition, the effect of space-charge-limited current flow upon the three-terminal characteristics of a device which is operating under *punch-through* conditions is also discussed.

73

4.2 OPERATION OF MOS TRANSISTORS AT VALUES OF APPLIED DRAIN VOLTAGE BELOW SATURATION

A cross-sectional representation of an n-channel MOSFET operating with a relatively low value of applied drain-to-source voltage is illustrated in Figure 4.1. Because the potential difference existing between the gate and drain electrodes is slightly less than the potential difference between the gate and the source, the electric field intensity in the gate insulator in the x direction will decrease with increasing y from the source end of the channel, $y = 0$, to the drain end of the channel, $y = L$. Consequently, as shown in the figure, the differential conductance will be greater near the source end of the channel than near the drain. The *differential resistance* of an element of the channel of length dy and width W is equal to

$$dR = -\frac{dy}{Q_n(y)\mu_n W},$$ (4.1)

where $Q_n(y)$ is the charge density in the n-type surface inversion layer per unit area as a function of y. The $Q_n(y)$ is a negative quantity. With the source and the substrate both kept at zero potential and with the applied drain voltage positive, the potential difference across the channel element is

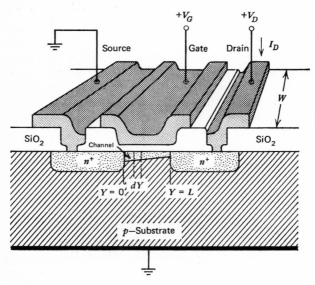

FIGURE 4.1   Cross-sectional view of an n-channel MOSFET operating below pinch-off.

equal to

$$dV = I_D \, dR = -\frac{I_D \, dy}{Q_n(y)\mu_n W}. \tag{4.2}$$

Integration over the channel from $y = 0$ (where $V = 0$) to $y = L$ (where $V = V_D$) yields an expression for the drain current:

$$I_D = -\frac{\displaystyle\int_{V=0}^{V=V_D} Q_n(y)\mu_n W \, dV}{\displaystyle\int_{y=0}^{y=L} dy}. \tag{4.3}$$

It can be assumed that the charge neutrality condition, as stated in (3.4), which relates the charge densities present on the gate electrode, within the gate insulator, and at the surface of the silicon, will be valid at any point in the structure as a function of y. The total charge density in the silicon surface per unit area, as given by (2.18), can be written as

$$Q_{\text{total}}(y) = Q_n(y) + Q_{SD_{\text{max}}}(y). \tag{4.4}$$

The $Q_{\text{total}}(y)$ can be expressed in terms of the gate voltage minus the flat-band voltage and the surface potential through the use of (3.6). The result is

$$Q_{\text{total}}(y) = C_{ox}[\phi_s(y) - (V_G - V_{FB})]. \tag{4.5}$$

It should be stressed that the charge neutrality condition was originally obtained from Gauss' law under the assumption that the electric field lines within the gate insulator were perpendicular to the surface of the silicon. (This was a natural consequence of the one-dimensional model which was employed.) However, in the structure of the MOSFET, the application of the drain-to-source potential will result in a transverse electric field in the y direction at the silicon surface. Thus for the original assumption that charge neutrality will exist at any point in the structure for $0 \leq y \leq L$ to hold, one must also make the assumption that the channel conductance varies gradually in the y direction and that the magnitude of the transverse field $\mathcal{E}_y$ is very much smaller than the perpendicular electric field $\mathcal{E}_x$. Consequently, the electric field lines present at the oxide-silicon interface are assumed to be *nearly* perpendicular to the surface of the silicon. This assumption is known as the *gradual channel approximation* and was first stated by Shockley in his treatment of the unipolar junction-gate field-effect transistor.[1]

If the source and the substrate are both kept at zero potential and the applied drain voltage is positive, the surface potential will increase with increasing values of y, as will the amount of reverse-bias across the channel-substrate junction. Under conditions such that strong inversion occurs at the silicon surface throughout the channel, the surface potential as a function of y will approximately be given by

$$\phi_s(y) \cong 2\phi_F + V(y), \tag{4.6}$$

where ϕ_F is the Fermi potential deep in the bulk of the p-type silicon substrate and $V(y)$ is the potential difference existing across the channel-substrate junction as a function of y.

The maximum width of the surface depletion region when a surface inversion layer is present, referring to (2.31), can be written as

$$x_{d_{\max}}(y) = \left[\frac{2\epsilon_s \phi_s(y)}{qN_A}\right]^{1/2}. \tag{4.7}$$

Note that the maximum width of the surface depletion region will increase with position from the source end of the channel to the drain end of the channel because of the previously mentioned reverse-bias across the channel-substrate junction. Use of (4.6) and (4.7) yields an expression for the charge in the depletion region per unit area for $0 \leq y \leq L$:

$$Q_{SD_{\max}}(y) = -qN_A x_{d_{\max}}(y) = -\{2\epsilon_s qN_A[2\phi_F + V(y)]\}^{1/2}. \tag{4.8}$$

The mobile charge density present in the n-type surface inversion layer can be obtained simply by combining (4.4) and (4.5):

$$Q_n(y) = -C_{ox}[(V_G - V_{FB}) - \phi_s(y)] - Q_{SD_{\max}}(y). \tag{4.9}$$

Substitution of (4.6) and (4.8) into (4.9) then yields

$$Q_n(y) = -C_{ox}[(V_G - V_{FB}) - V(y) - 2\phi_F] + \{2\epsilon_s qN_A[2\phi_F + V(y)]\}^{1/2}. \tag{4.10}$$

The drain current may now be calculated by substituting (4.10) into (4.3) and integrating. The result is

$$I_D = \frac{+\mu_n W C_{ox}\{V_D[V_G - V_{FB} - 2\phi_F - (V_D/2)] - (2(2\epsilon_s qN_A)^{1/2}/3C_{ox})[(V_D + 2\phi_F)^{3/2} - (2\phi_F)^{3/2}]\}}{L}. \tag{4.11}$$

Now, defining

$$\beta \equiv \frac{\epsilon_{ox}\mu_n W}{T_{ox}L} = \frac{\mu_n W C_{ox}}{L} \tag{4.12}$$

and

$$\Phi \equiv \frac{T_{ox}}{\epsilon_{ox}}(2\epsilon_s qN_A)^{1/2} = \frac{(2\epsilon_s qN_A)^{1/2}}{C_{ox}}, \tag{4.13}$$

the drain current can be written as

$$I_D = \beta \left\{ V_D \left(V_G - V_{FB} - 2\phi_F - \frac{V_D}{2} \right) - \tfrac{2}{3}\Phi[(V_D + 2\phi_F)^{3/2} - (2\phi_F)^{3/2}] \right\}.$$

$$(4.14)$$

The dependence of the function Φ upon the doping concentration in the silicon substrate is shown in the graph of Figure 4.2.

4.2.1   Drain-to-Source Conduction at Very Low Drain Voltages

Equation 4.14 describes the operation of an n-channel MOSFET for *all* drain voltages below the onset of current saturation. A similar expression for the drain current of a p-channel MOSFET operating below current saturation can be obtained using an analogous procedure.

At very low values of drain-to-source voltage such that

$$V_D \ll V_G - V_{FB} - 2\phi_F \qquad (4.15)$$

and

$$V_D \ll 2\phi_F, \qquad (4.16)$$

it follows that

$$(V_D + 2\phi_F)^{3/2} \rightarrow (2\phi_F)^{3/2} + \tfrac{3}{2}(2\phi_F)^{1/2}V_D, \qquad (4.17)$$

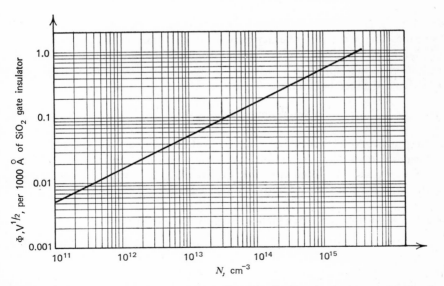

FIGURE 4.2   Variation of the function Φ with effective doping concentration in the silicon substrate.

and, consequently, the current-voltage relationship for very small values of applied drain-to-source potential is approximately equal to

$$I_D \cong \beta V_D [V_G - V_{FB} - 2\phi_F - \Phi(2\phi_F)^{1/2}]. \tag{4.18}$$

It is interesting to note that

$$-\Phi(2\phi_F)^{1/2} = \frac{T_{ox}}{\epsilon_{ox}} [2\epsilon_s q N_A (2\phi_F)]^{1/2} = Q_{SD_{max}} \left(\frac{T_{ox}}{\epsilon_{ox}}\right). \tag{4.19}$$

That is, the charge in the depletion region per unit area under strong inversion conditions is given by $-C_{ox}(2\phi_F)^{1/2}\Phi$. Remembering that the flatband voltage is defined as

$$V_{FB} = \phi_{MS'} - \frac{Q_{SS}}{C_{ox}}, \tag{3.3}$$

and the threshold voltage is given by

$$V_T = \left(\frac{-Q_{SS} - Q_{SD_{max}}}{\epsilon_{ox}}\right) T_{ox} + \phi_{MS'} + 2\phi_F, \tag{2.44}$$

then it can be easily seen through the use of (4.18), (4.19), (3.3), and (2.44) that the drain current can be written approximately as

$$I_D \cong \beta V_D (V_G - V_T) \tag{4.20}$$

for very small values of applied drain-to-source voltage. Equation 4.20 describes the electrical characteristics of an *ideal linear resistor* whose (variable) resistance is inversely proportional to the gate voltage minus the threshold voltage. In other words, in the limit of very low drain voltages, the effective resistance between the drain and source electrodes approaches

$$\lim_{V_D \to 0} R_{DS} \cong [\beta(V_G - V_T)]^{-1}. \tag{4.21}$$

In practice, (4.21) will be valid for gate-to-source voltages which are sufficiently low such that the mobility of the electrons in the channel remains fairly constant, and when the effects of any parasitic resistance associated with the drain and source electrodes are negligible.

4.2.2   Drain-to-Source Conduction Slightly Below Saturation

As the drain-to-source voltage is increased with the applied gate voltage held constant, the differential conductance gradually decreases and the observed drain current begins to saturate. Consequently, (4.20) and (4.21) will no longer hold and the linearity of the variable resistance relationship will begin to deteriorate. Under these conditions, the approximations made for the case of very low drain-to-source voltages are no longer valid and (4.14), in its entirety, must be used to calculate the drain current. Equation 4.14 can be written as

$$I_D = k(V_G)V_D - \Delta(V_D), \tag{4.22}$$

where the parameters k and Δ are defined as

$$k(V_G) \equiv \beta(V_G - V_{FB} - 2\phi_F), \tag{4.23}$$

and

$$\Delta(V_D) = \tfrac{1}{2}\beta V_D^2 + \frac{2\beta\Phi}{3}[(V_D + 2\phi_F)^{3/2} - (2\phi_F)^{3/2}]. \tag{4.24}$$

The Δ can be considered to be the deviation from linearity when the MOSFET is operating as a *gate-voltage-controlled* variable resistor. For increasing drain-to-source voltage, the effect of the Δ correction factor becomes greater and the slope of the drain current-versus-drain voltage curve decreases until current saturation takes place.

The electrical characteristics of a typical *p*-channel enhancement-type MOSFET operating at low drain voltages are shown in the photographs of Figure 4.3. As can be seen in Figure 4.3*a*, the device approximates a voltage-variable resistor reasonably well for small values of both positive and negative drain voltage. However, as can be seen in Figure 4.3*b*, the dynamic range of the linear variable resistance region is limited for negative applied drain voltage by the onset of current saturation, and for positive applied drain voltage by the turn-on of the drain-to-substrate diode.

Ideally, when current saturation occurs, the slope of the drain current-versus-drain voltage curve should go to zero. In practice, however, the observed slope will be slightly greater than zero because of second-order effects that result in a finite dynamic drain resistance in the region of saturated current flow. These effects are discussed in detail in later sections. Assuming, for the time being, that these second-order effects can be neglected, the drain voltage at which current saturation will be observed for a given value of gate voltage can be obtained by setting the derivative of the drain current with respect to the drain voltage equal to zero. Differentiating (4.14) yields

$$\left(\frac{\partial I_D}{\partial V_D}\right) = \beta[V_G - V_{FB} - 2\phi_F - V_D - \Phi(V_D + 2\phi_F)^{1/2}] = 0. \tag{4.25}$$

Solving (4.25) yields the value of drain voltage at which current saturation will occur as a function of the gate voltage. Denoting this value of drain voltage by $V_{D\text{sat}}$, the result is

$$V_{D\text{sat}} = V_G - V_{FB} - 2\phi_F + \frac{\Phi^2}{2} - \Phi\left(V_G - V_{FB} + \frac{\Phi^2}{4}\right)^{1/2}. \tag{4.26}$$

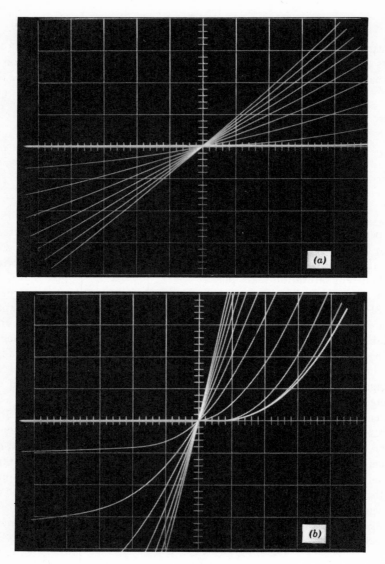

FIGURE 4.3 Electrical characteristics of a Motorola 2N4352 p-channel enhancement type MOSET; (a) vertical scale, drain current: $-500 \mu A/\text{div.}$; horizontal scale, drain voltage: -0.1 V/div.; gate voltage: 0 to -10 V, in -1 V steps; substrate and source held at 0 V; (b) vertical scale, drain current: $-500 \mu A/\text{div.}$; horizontal scale, drain voltage: -0.5 V/div.; gate voltage: 0 to -10 V, in -1 V steps; substrate and source held at 0 V.

Now, since

$$\Phi\left(V_G - V_{FB} + \frac{\Phi^2}{4}\right)^{1/2} > \frac{\Phi^2}{2} \tag{4.27}$$

for all values of gate voltage greater than the flat-band voltage, it follows that the drain current will saturate at lower values of drain voltage (for a given value of gate voltage) as the magnitude of Φ becomes larger with increased substrate doping. Thus, for MOSFETs fabricated on low-resistivity substrates (where Φ is comparatively large) current saturation will occur at lower drain voltages and the "knee" of the drain current-versus-drain voltage curve will move closer to the origin. The MOSFETs fabricated on relatively high resistivity substrates, on the other hand, will exhibit drain current saturation at the (higher) drain voltages predicted by first-order theory, since the amount of charge density contained in the surface depletion region will be comparatively small and Φ will be very much less than one. It can easily be shown that for such a device fabricated on a high-resistivity substrate, the drain voltage at which current saturation occurs will approximately be given by

$$V_{D\text{sat}} \cong V_G - V_T. \tag{4.28}$$

4.3 CONDUCTION IN THE REGION OF SATURATED DRAIN CURRENT

In the previous section, it was demonstrated that as the drain voltage is increased from zero with the gate-to-source potential held constant, the drain-to-source differential conductance gradually decreases until current saturation takes place. To first order, the drain-to-source current becomes independent of the applied drain voltage for all drain voltages greater than $V_{D\text{sat}}$ and less than the breakdown voltage of the drain diode. The physical basis for the saturation of the drain-to-source current can be easily understood by referring to Figure 4.4. As shown in Figure 4.4a, for the case of an n-channel MOSFET fabricated on a moderately high-resistivity substrate with $\Phi \ll 1$, with the application of a small positive drain voltage $(+V_{D1})$ the induced channel will extend throughout the region between drain and source as long as

$$V_{D1} \ll (V_G - V_T). \tag{4.29}$$

If the potential at any point in the channel is denoted by $V(y)$, where

$$0 \leq V(y) \leq V_D, \tag{4.30}$$

and if all the gate voltage is assumed to be impressed across the gate insulator, then the perpendicular electric field in the insulator as a function of y will be

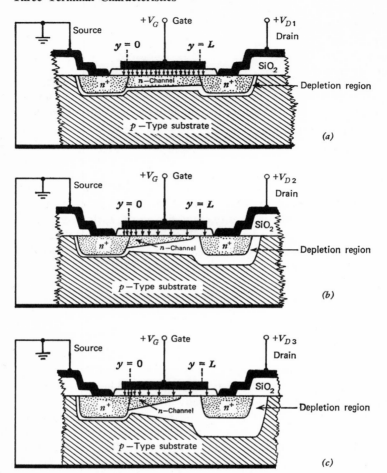

FIGURE 4.4 The onset of drain current saturation with increasing drain voltage in an *n*-channel MOSFET; (*a*) below pinch-off; (*b*) at pinch-off; (*c*) for drain voltages greater than $V_{D\text{sat}}$.

given by

$$\mathscr{E}_{ox} = \frac{V_G - V(y)}{T_{ox}}.$$ (4.31)

The boundary conditions at the drain and source, respectively, are

$$V(y = L) = V_D$$ (4.32)

and

$$V(y = 0) = 0.$$ (4.33)

Therefore, when the applied drain voltage is equal to $+V_{D1}$, it is clear that the electric field in the gate insulator, as given by (4.31), will be greater near the source than it is near the drain. It follows that the density of negative surface charge that is required to terminate this field will vary in the same manner as a function of y. Consequently, the total electron carrier concentration in the inversion layer will be largest near the source region and will decrease steadily with increasing y.

As the magnitude of the drain voltage increases, the variation of the electric field in the oxide as a function of y will become more pronounced and the free carrier concentration in the channel near the drain electrode will decrease. When the drain voltage is increased to a value $+V_{D2}$ such that

$$V_{D2} \approx (V_G - V_T), \tag{4.34}$$

the electric field in the gate oxide at $y = L$, from (4.31), will be equal to

$$\mathscr{E}_{ox}(y = L) \cong \frac{V_T}{T_{ox}}. \tag{4.35}$$

The electron concentration in the channel, near the source, will remain unchanged since the source is held at zero potential and the electric field in the oxide at $y = 0$ will not vary with changes in the applied drain voltage. From the definition of the threshold voltage, (4.35) requires that at $y = L$:

1. The mobile (electronic) charge density must go to zero.
2. The electric field in the oxide must be terminated with depletion charge only.

Thus as illustrated in Figure 4.4b, with $V_D = +V_{D2}$, a surface depletion region will form at $y = L$ and block or "pinch-off" the drain end of the channel. The length of the depletion region that forms at pinch-off will be much smaller than the length of the channel. For drain-to-source voltages above pinch-off, the voltage drop across the channel will remain practically constant and any additional voltage will be impressed across the depletion region, which will widen and spread toward the source, as shown in Figure 4.4c. As a result of the widening of the drain depletion region, the effective channel length will decrease. However, for typical MOSFET structures, the reduction in the effective length of the channel will be quite small compared to the drain-to-source spacing, L. To first order, the effective channel length can be assumed to be a constant independent of the applied drain voltage. Consequently, with constant gate voltage, the resistance of the channel will remain constant. Since both the voltage drop across the channel and the resistance of the channel will remain practically constant with increasing drain voltage, it follows that the drain current beyond pinch-off will also remain fairly constant. In other words, current saturation will be observed

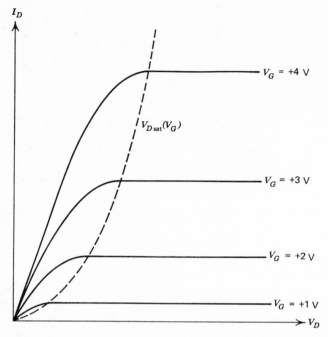

FIGURE 4.5 Saturation of the drain current as a function of the applied gate-to-source voltage for an *n*-channel MOSFET.

for all drain voltages greater than $+V_{D2} = V_{Dsat}$, where V_{Dsat} is given by (4.26). As can be seen from the equation, the value of drain voltage at which current saturation occurs will increase with increasing gate voltage. The behavior of V_{Dsat} as a function of V_G for a typical *n*-channel MOSFET is illustrated in Figure 4.5.

4.3.1 The Saturated Drain Current as a Function of the Gate Voltage

The saturated drain-to-source current, I_{Dsat}, can be expressed as a function of the applied gate-to-source voltage by substituting (4.26) into (4.14). The result is

$$I_{Dsat} = \beta\{V_{Dsat}(V_G - V_{FB} - 2\phi_F - \tfrac{1}{2}V_{Dsat})$$
$$- \tfrac{2}{3}\Phi[(V_{Dsat} + 2\phi_F)^{3/2} - (2\phi_F)^{3/2}]\}. \quad (4.36)$$

The V_{Dsat} can be expressed as

$$V_{Dsat} = V_G - V_{FB} - 2\phi_F + F(\Phi, V_G), \quad (4.37)$$

where the function $F(\Phi, V_G)$ is defined as

$$F(\Phi, V_G) \equiv \frac{\Phi^2}{2} - \Phi\left(V_G - V_{FB} + \frac{\Phi^2}{4}\right)^{1/2}. \qquad (4.38)$$

The dependence of $F(\Phi, V_G)$ is plotted as a function of Φ in Figure 4.6 for three different values of the gate voltage minus the flat-band voltage. [It is evident from inspection of (4.27) that $F(\Phi, V_G)$ will always be less than or equal to zero.] Combining (4.36) and (4.37) yields

$$I_{Dsat} = \beta\{[\tfrac{1}{2}(V_G - V_{FB} - 2\phi_F + F(\Phi, V_G))][V_G - V_{FB} - 2\phi_F - F(\Phi, V_G)]$$
$$- \tfrac{2}{3}\Phi\{[V_G - V_{FB} + F(\Phi, V_G)]^{3/2} - (2\phi_F)^{3/2}\}\}, \qquad (4.39)$$

or

$$I_{Dsat} = \frac{\beta}{2}\{(V_G - V_{FB} - 2\phi_F)^2 - F^2(\Phi, V_G)$$
$$- \tfrac{4}{3}\Phi\{[V_G - V_{FB} + F(\Phi, V_G)]^{3/2} - (2\phi_F)^{3/2}\}\}. \qquad (4.40)$$

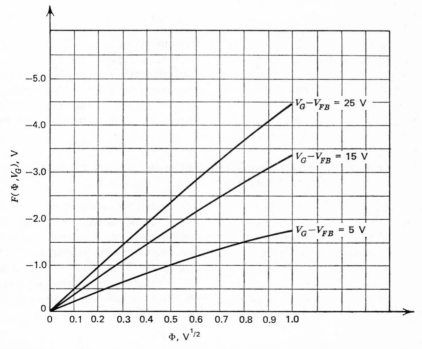

FIGURE 4.6 Behavior of the function $F(\Phi, V_G)$ versus Φ for different values of $V_G - V_{FB}$.

It is interesting to consider the limiting form of the general expression for the saturated drain current, as given by (4.40), for an n-channel MOSFET fabricated on a moderately high-resistivity p-type substrate with $\Phi \ll 1$ and applied gate voltage sufficiently high such that $(V_G - V_{FB} - 2\phi_F) \gg 1$. Under these conditions, since the effect of the charge density in the surface depletion region is virtually negligible, it can be shown that the saturated drain current as given by (4.40) will be approximately equal to

$$I_{D\text{sat}} \cong \frac{\beta}{2}(V_G - V_T)^2. \tag{4.41}$$

Equation 4.41 expresses the well-known "square-law" dependence of the drain current in the saturation region upon the gate-to-source voltage for an n-channel device. The equation is valid for all drain voltages greater than the gate voltage minus the threshold voltage and less than the breakdown voltage of the drain diode. Similarly, for a p-channel MOSFET fabricated on a moderately high-resistivity substrate, the saturated drain current will be approximately given by

$$I_{D\text{sat}} = -\frac{\beta}{2}(V_G - V_T)^2. \tag{4.42}$$

It should be noted that the approximations given in (4.41) and (4.42) were derived under the assumption of complete current saturation and, consequently, are independent of the applied drain voltage. In practice, however, the experimentally observed drain current for MOSFETs operating beyond pinch-off will slowly increase with increasing drain voltage. The second-order effects that result in incomplete current saturation will be treated in the following sections.

4.3.2   Incomplete Current Saturation: Conduction Beyond Pinch-Off as a Function of the Applied Drain-to-Source Voltage

In general, complete saturation of the drain current for values of the applied drain voltage which are greater than the pinch-off voltage will only be observed in MOSFETs with very large channel lengths. As the drain-to-source spacing, L, of a MOSFET is reduced, the saturation properties of the drain current beyond pinch-off rapidly deteriorate. This is particularly true of MOS devices fabricated on high-resistivity substrates. Three separate mechanisms can contribute to the degradation of current saturation in MOSFETs operating beyond pinch-off:

1. Modulation of the effective length of the channel by the spreading of the drain depletion region.

2. Electrostatic feedback of the drain field into the channel.

3. Space-charge-limited current flow between the drain and source regions after the drain depletion region has spread and punched-through to the source.

4.3.3 Modulation of the Effective Channel Length by the Spreading of the Drain Depletion Region

As previously stated, for applied drain voltages beyond pinch-off, a depletion region will form at the end of the channel near the drain diffusion and spread toward the source diffusion with increasing V_D. As long as the drain-to-source spacing is much greater than the width of the drain depletion region, the overall reduction in the length of the conducting portion of the channel will be quite small and, to first order, the saturated drain current will be constant and independent of the drain voltage. However, when the drain-to-source spacing is small enough so as to be comparable to the width of the drain depletion region at a given value of drain voltage, the reduction in the effective channel length with increasing V_D will be substantial. As the effective length of the conducting portion of the channel decreases, its resistance will also decrease and, as a result, the observed drain current beyond pinch-off will increase with increasing drain voltage. Thus complete current saturation will not be observed. This is illustrated in Figure 4.7, which shows the electrical characteristics of a typical n-channel MOSFET with a small drain-to-source spacing. The incomplete current saturation observed in this device is almost entirely a result of the modulation of the effective channel length by the spreading of the drain depletion region.

When the mechanism above is dominant, the behavior of the drain current beyond pinch-off can be studied by considering the spreading of the drain depletion region as a function of the applied drain-to-source voltage. Referring to the cross-sectional view of an n-channel MOSFET shown in Figure 4.8, if L denotes the drain-to-source spacing and ΔL denotes the width of the depletion region near the drain end of the channel, then for drain voltages such that $V_D \geqq V_{D\text{sat}}$, the drain current will be given by

$$I_{D\text{sat}'} \cong \frac{I_{D\text{sat}}}{[1 - (\Delta L / L)]}, \tag{4.43}$$

where $I_{D\text{sat}}$ is the drain current under conditions of complete current saturation, as predicted by first-order theory. The width of the depletion region at the drain end of the channel near the silicon surface will be a function of the drain voltage and will also be significantly influenced by the presence of the overlying gate electrode.[2] The ΔL will also depend on the applied gate voltage, the thickness of the insulating gate oxide, and the magnitude of the

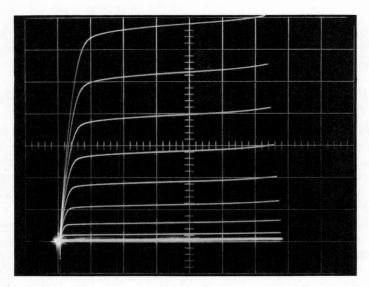

FIGURE 4.7 Three-terminal characteristics of a Motorola 2N4351 n-channel enhancement-type MOSFET fabricated on a low-resistivity p-type substrate; vertical scale, drain current: $+1$ mA/div.; horizontal scale, drain voltage: $+2$ V/div.; gate voltage: 0 to $+6$ V, in $+\frac{1}{2}$ V steps, substrate and source held at 0 V.

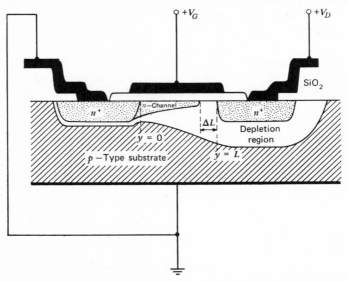

FIGURE 4.8 Cross-sectional view of an n-channel MOSFET operating beyond pinch-off, in which the conductance in the saturated drain current region is primarily determined by the spreading of the drain depletion region into the channel.

fixed positive charge density per unit area located at the oxide-silicon interface, Q_{SS}. However, if these effects are neglected and the drain-substrate junction is assumed to be abrupt, then from simple *p-n* junction theory, with a voltage of $(V_D - V_{D\text{sat}})$ applied across the depletion region, ΔL can be approximated by

$$\Delta L \cong \left[\frac{2\epsilon_s(V_D - V_{D\text{sat}})}{qN_A} \right]^{1/2}. \tag{4.44}$$

The *conductance* in the region of saturated drain current can be defined as

$$g_{D\text{sat}} \equiv \left(\frac{\partial I_{D\text{sat'}}}{\partial V_D} \right). \tag{4.45}$$

Similarly, the *drain saturation resistance* will be defined by

$$r_{D\text{sat}} \equiv \frac{1}{g_{D\text{sat}}} = \left(\frac{\partial I_{D\text{sat'}}}{\partial V_D} \right)^{-1}. \tag{4.46}$$

Combining (4.43) and (4.45) yields

$$g_{D\text{sat}} = I_{D\text{sat}} \frac{\partial}{\partial V_D} \left(1 - \frac{\Delta L}{L} \right)^{-1} = I_{D\text{sat}} \frac{\partial}{\partial V_D} \left(\frac{L}{L - \Delta L} \right). \tag{4.47}$$

Substituting (4.44) into (4.47) and differentiating gives the following expression for the conductance in the region of saturated drain current:

$$g_{D\text{sat}} = \frac{I_{D\text{sat}}K_s L}{2[L - K_s(V_D - V_{D\text{sat}})^{1/2}]^2 (V_D - V_{D\text{sat}})^{1/2}}, \tag{4.48}$$

where the constant K_s is defined as

$$K_s \equiv \left(\frac{2\epsilon_s}{qN_A} \right)^{1/2}, \tag{4.49}$$

and $V_{D\text{sat}}$ is given by (4.26). It is interesting to note that

$$\Delta L = K_s(V_D - V_{D\text{sat}})^{1/2}, \tag{4.50}$$

and, consequently, $g_{D\text{sat}}$ can be expressed as

$$g_{D\text{sat}} = \frac{I_{D\text{sat}}K_s L}{2(L - \Delta L)^2(V_D - V_{D\text{sat}})^{1/2}}, \tag{4.51}$$

where $I_{D\text{sat}}$, in its general form, is given by (4.40). Equation 4.51 can be easily rearranged to yield

$$g_{D\text{sat}} = \frac{I_{D\text{sat}}K_s}{2L[1 - (\Delta L/L)]^2(V_D - V_{D\text{sat}})^{1/2}} = \frac{1}{r_{D\text{sat}}}. \tag{4.52}$$

It can easily be seen from (4.52) that as the drain-to-source spacing associated with a MOSFET is made smaller so as to become comparable to ΔL, the fraction of the total channel length that is modulated by the spreading of the drain depletion region will become much greater; hence the drain saturation resistance will decrease rapidly.

4.3.4 Electrostatic Feedback of the Drain Field into the Channel Region

Modulation of the effective length of the conducting channel as a result of the spreading of the depletion region near the drain end of the channel has been shown to be the dominant mechanism contributing to the reduction of drain saturation resistance with decreasing drain-to-source spacing for MOSFETs fabricated on low-resistivity substrates.[3] However, for devices fabricated on moderately high-resistivity substrates, another mechanism has been found to dominate. This mechanism adds to the effect of channel-length-modulation to produce a further degradation in the drain saturation resistance.[3,4] For this reason, MOSFETs with small channel lengths fabricated on lightly doped substrates will usually exhibit much lower values of drain saturation resistance when compared with similar devices fabricated on highly doped substrates with all other device parameters held constant.

The physical basis for this second mechanism can be seen by referring to Figure 4.9, which illustrates a cross-sectional representation of an n-channel MOSFET fabricated on a moderately high-resistivity p-type silicon substrate. The device is operating beyond pinch-off, in the region of saturated drain current. As can be seen from the figure, the widths of the depletion regions at the drain-substrate and channel-substrate junctions can become comparable to the drain-to-source spacing of the device if the resistivity is sufficiently high. Under these conditions, an appreciable amount of capacitive coupling can occur from the drain electrode to the channel region. As illustrated in Figure 4.9, electric field lines originating at the drain can extend through the deep depletion region and terminate in the channel. As the drain voltage is increased, the electric field intensity in this region also increases and the electron population in the n-type inversion layer must increase to completely terminate the larger field. Thus the drain electrode is actually acting as a "second gate" in controlling the drain-to-source conductance. A low-resistivity substrate, on the other hand, will act as an electrostatic shield since the much narrower depletion region width effectively decouples the drain field from the channel. This results in significantly higher observed values of drain saturation resistance with low-resistivity substrates for similar values of channel length.

When electrostatic feedback of the drain field into the channel region is the dominant mechanism leading to incomplete saturation of the drain-to-source

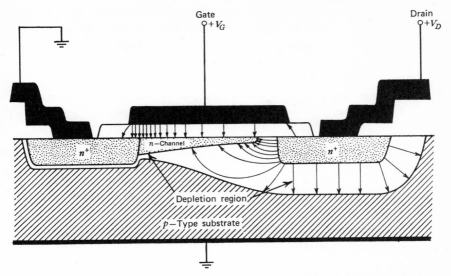

FIGURE 4.9 Cross-sectional view of an *n*-channel MOSFET operating beyond pinch-off, in which the electrostatic feedback of the drain field into the channel is the dominant mechanism in determining the conductance in the saturated drain current region.

current, the drain saturation resistance will again be observed to decrease as the drain-to-source spacing of the device is made smaller. This is because a greater portion of the conducting channel region will come under the influence of the drain field with decreasing values of L.

The electrostatic feedback of the drain field into the channel and its effect on the saturated drain current characteristics of MOSFETs was first discussed in a semiquantitative treatment by Hofstein and Heiman.[4] Following their analysis, the mechanism can be studied by considering the action of the drain electrode as a second gate as it couples through the underlying depletion region into the channel. If the voltage applied to the drain electrode of an *n*-channel MOSFET is increased by an amount ΔV_D, the corresponding change in the *average* charge density in the conducting portion of the channel *per unit area* due to the coupling field can be expressed as

$$Q_{av} = -\frac{C_{dct}\Delta V_D}{WL} \qquad (4.53)$$

for values of gate voltage much greater than the threshold voltage. The C_{dct} is the *total* effective coupling capacitance between the drain and the conducting channel. For drain voltages beyond pinch-off, the voltage impressed across the conducting portion of the channel will be approximately equal to

$V_G - V_T$. Under conditions such that $\Delta L \ll L$, the average electric field in the y direction in the channel will approximately be equal to

$$\mathscr{E}_{av} = -\frac{(V_G - V_T)}{L}, \tag{4.54}$$

and the average drift velocity of the electrons in the conducting portion of the channel will be given by

$$v_{d_{av}} = +\frac{\mu_n(V_G - V_T)}{L}. \tag{4.55}$$

Therefore, when the drain voltage is increased by an amount ΔV_D, the corresponding change in the total drain current due to electrostatic feedback can be written as

$$\Delta I_{Dsat'} \cong \frac{\mu_n C_{dct} \Delta V_D (V_G - V_T)}{L^2}. \tag{4.56}$$

Rearranging (4.56) yields the following expression for the conductance in the region of saturated drain current when the electrostatic feedback mechanism is dominant:

$$g_{Dsat} \equiv \frac{\partial I_{Dsat'}}{\partial V_D} \cong \frac{\mu_n C_{dct}(V_G - V_T)}{L^2}. \tag{4.57}$$

Under normal operating conditions, C_{dct} will be a slowly varying function of the drain voltage.[4] Hence for substrate resistivities sufficiently high so that the effect of the electrostatic feedback of the drain field into the channel is much greater than the effect of the modulation of the effective channel length by the spreading of the drain depletion region, g_{Dsat} will be observed to be proportional to $V_g - V_T$. This can be seen in the experimental data shown in Figure 4.10, derived from a p-channel MOSFET fabricated on a medium-resistivity (21 Ω-cm n-type) silicon substrate. The *actual* spacing between the drain and source regions of the device was approximately 0.35 mil, after the effects of lateral diffusion. The slope of the graph indicates that $I_{Dsat'}$ is directly proportional to the square of g_{Dsat}. Since $I_{Dsat'}$ is itself proportional to $(V_G - V_T)^2$, it follows that, for this device, g_{Dsat} is directly proportional to the gate voltage minus the threshold voltage, as predicted by (4.57).

As previously stated, the drain electrode can actually act as a second gate by modulating the conductance of the channel region with the coupling electric field. It is particularly interesting to note the relative effects of both the drain voltage *and* the gate voltage on the drain-to-source current beyond

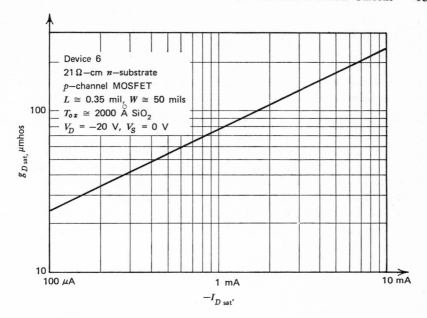

FIGURE 4.10 Experimental data relating $g_{D\text{sat}}$ to $I_{D\text{sat}'}$ in a p-channel MOSFET in which the electrostatic feedback mechanism is dominant.

pinch-off and to compare $g_{D\text{sat}}$ with g_m, the transconductance. The transconductance, which can be obtained by differentiating (4.41) with respect to the gate voltage, is approximately given by

$$g_m \cong \frac{\partial I_{D\text{sat}}}{\partial V_G} \cong \beta(V_G - V_T) = \frac{\mu_n W C_{ox}(V_G - V_T)}{L}. \qquad (4.58)$$

Since both $g_{D\text{sat}}$ and g_m are proportional to the gate voltage minus the threshold voltage, their ratio is simply

$$\frac{g_{D\text{sat}}}{g_m} \cong \frac{C_{dct}}{C_{ox}WL}. \qquad (4.59)$$

The total oxide capacitance, which will be denoted by C_{tox}, is related to the oxide capacitance per unit area by

$$C_{tox} = C_{ox}WL; \qquad (4.60)$$

therefore, it follows that the ratio of $g_{D\text{sat}}$ to g_m is equal to a simple ratio of capacitances:

$$\frac{g_{D\text{sat}}}{g_m} \cong \frac{C_{dct}}{C_{tox}}. \qquad (4.61)$$

In general, the total oxide capacitance will be much greater than the total drain-to-channel coupling capacitance for a typical device; thus $g_{D\text{sat}}$ will usually be much less than g_m.

4.3.5   Space-Charge-Limited Current Flow from Drain to Source after Punch-Through of the Drain Depletion Region Has Occurred

If the spacing between the drain and source regions of a MOSFET is sufficiently small and the resistivity of the substrate on which the device is fabricated is sufficiently high, the depletion region associated with the reverse-biased drain-substrate junction can spread with increasing drain voltage and will eventually touch the depletion region associated with the source-substrate junction. This condition is commonly referred to as *punch-through*, and it is illustrated in the cross-sectional view of a silicon *n*-channel MOSFET shown in Figure 4.11. For increasingly positive applied drain voltage beyond

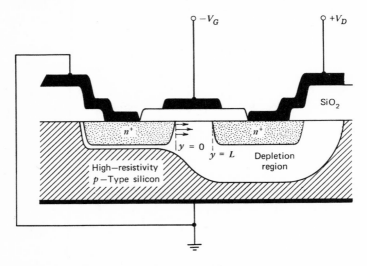

FIGURE 4.11   Cross-sectional representation of an *n*-channel MOSFET in which punch-through has taken place.

punch-through, majority carriers in the source region, electrons in this case, can be injected into the depleted channel region where they will be swept by the high transverse electric field and will be collected at the drain. If the drain-substrate diode is assumed to be an ideal step junction, the drain voltage required to achieve punch-through for an *n*-channel MOSFET fabricated on a high-resistivity *p*-type substrate with drain-to-source spacing

L will be approximately given by

$$V_{pt} \cong \frac{qN_AL^2}{2\epsilon_s} - V_o, \qquad (4.62)$$

where V_o is the contact potential of the drain-substrate diode, and the width of the depletion region associated with the drain-substrate junction is assumed to be much greater than the width of the depletion region associated with the source-substrate junction. As can be seen from (4.62), for devices with extremely small channel lengths that are fabricated on very high-resistivity substrates, it is therefore possible for punch-through to occur with the applied drain-to-source voltage equal to zero.

It should be noted that (4.62) does not take into account any surface effects on the spreading of the drain depletion region through the channel region toward the source. As previously mentioned in Section 4.3.3, the width of the depletion region near the silicon surface for a given value of applied drain voltage will be influenced by the presence of the insulated gate structure and will be a function of the gate voltage, the oxide thickness, and Q_{SS}.

To study the nature of the drain-to-source conduction mechanism in a MOSFET for drain voltages beyond punch-through, it is interesting to neglect the effect of the insulated gate electrode by treating the simplified gap-type semiconductor structure shown in Figure 4.12. The structure consists of two parallel n^+ regions of area A which have been diffused into opposite sides of a high-resistivity p-type (π) silicon substrate and are separated by a very small distance, L. As a further simplification, it will be assumed that the distance L is sufficiently small, and the resistivity of the high-resistivity region which separates the two diffused regions is sufficiently high, so that punch-through has already occurred in the structure with the applied voltage $+V_D$ equal to zero. Thus the energy-band diagram for the structure under zero-bias conditions will be as shown in Figure 4.13a. Note that since the n^+ regions are doped to degeneracy, the Fermi level in these regions lies above the conduction band edge. Under punch-through conditions, the depletion region will extend throughout the gap from $y = 0$ to $y = L$. The depletion region will consist almost entirely of ionized acceptors and the free carrier concentration will be extremely small. Ordinarily, since the free carrier concentration in this region is very low, very little current would be expected to flow from "drain" to "source" with the application of a positive voltage, $+V_D$. However, since the distance L is quite small, the possibility of electron injection into the conduction band of the depletion region must be considered. The electrons can be injected from the highly doped "source" region into the depletion region by Schottky emission, and if L is sufficiently small, they can reach the "drain" electrode where they will be collected.

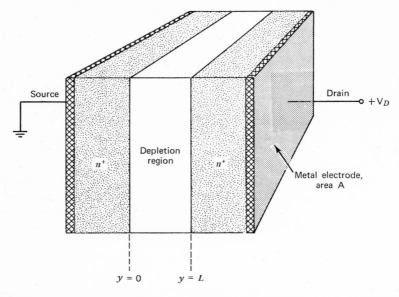

FIGURE 4.12 Simplified $n^+\pi n^+$ gap-type semiconductor structure with applied voltage.

Appreciable current flow will only be observed if the transit time of the injected electrons, t, is shorter than their relaxation time within the gap region, τ. That is,

$$t = \frac{L}{v_n} = \frac{L^2}{\mu_n V_D} < \tau, \qquad (4.63)$$

where v_n is the velocity of the injected electrons in the y direction, and μ_n is their mobility within the gap region. For very small values of L, it is also possible for electrons to penetrate the potential barrier and tunnel through the depletion region. In either case, the electrons that are injected into the depletion region will create a *large negative space-charge region* in the gap from $y = 0$ to $y = L$. The concentration of electrons will be largest near the injecting electrode at $y = 0$ and will decrease with increasing y across the gap region. Consequently, with the application of positive voltage across the structure, the energy band diagram under injection conditions will be as shown in Figure 4.13b. The presence of the large concentration of negative charge in the space charge region near $y = 0$ will tend to limit the amount of additional electron injection from the "source"; therefore, the total observed current across the gap region will become "space-charge-limited." Neglecting fringing effects, the total drain-to-source space-charge-limited current can be

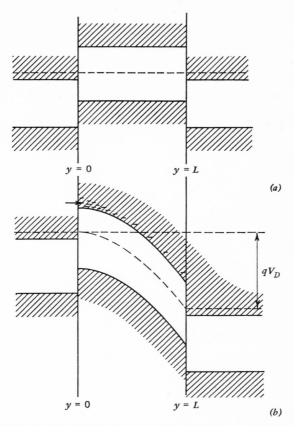

FIGURE 4.13 Energy band diagram associated with space-charge-limited injection of electrons across a gap-type semiconductor structure; (*a*) zero bias condition; (*b*) electron injection with applied positive bias.

assumed to be a drift current that can be expressed as

$$I_{SCL} = -nqv_nA. \tag{4.64}$$

The negative sign in (4.64) assures that the current will flow in the $-y$ direction for a positive applied drain voltage. Now n, the electron concentration per unit volume in the gap region, will be a function of y. Under steady state conditions, the current density in the y direction must be constant from $y = 0$ to $y = L$, and this requires that the product of the electron concentration times the electron velocity be constant within this region. Since n will be large near $y = 0$, the electron velocity there will be much less than near $y = L$, where n will be comparatively small. The velocity of electrons in the gap region, as a function of y, will be related to the electric field in the y direction by

$$v_n(y) = -\mu_n \mathscr{E}_y(y), \tag{4.65}$$

where $\mathscr{E}_y$ is directed in the negative y direction with positive applied drain voltage, $+V_D$. Both $-\mathscr{E}_y$ and v_n will be at a minimum at $y = 0$ and will increase with increasing y.

The electric field in the y direction within the gap region can be related to the electron concentration per unit volume through the use of Poisson's equation

$$\frac{d\mathscr{E}_y}{dy} = -\frac{(-qn)}{\epsilon_s}, \tag{4.66}$$

or

$$qn = \epsilon_s \frac{d\mathscr{E}_y}{dy}. \tag{4.67}$$

Substituting (4.67) into (4.64) yields

$$I_{SCL} = -\epsilon_s v_n A \frac{d\mathscr{E}_y}{dy} = \epsilon_s \mu_n A \mathscr{E}_y \frac{d\mathscr{E}_y}{dy}. \tag{4.68}$$

Equation 4.68 can be integrated and solved for the electric field intensity. The result is

$$\mathscr{E}_y = -\left[\frac{2(yI_{SCL} + C)}{\epsilon_s \mu_n A}\right]^{1/2}, \tag{4.69}$$

where C is a constant of integration. Since the magnitude of the electric field intensity is smallest at $y = 0$ due to the large electron concentration there, the constant C can be set equal to zero if the assumption is made that

$$\mathscr{E}_y(y = 0) \cong 0. \tag{4.70}$$

[It should be mentioned that although the magnitude of the electric field at $y = 0$ is much smaller than it is near the drain, it must be *non-zero* to maintain the flow of the drift current from the source. However, for drain voltages greater than a few volts, the assumption expressed by (4.70) is extremely good.] The drain voltage can now be obtained by integrating $\mathscr{E}_y$ from $y = 0$ to $y = L$. Thus

$$V_D \cong -\int_0^L \mathscr{E}_y \, dy = L^{3/2}\left(\frac{8I_{SCL}}{9\epsilon_s \mu_n A}\right)^{1/2}. \tag{4.71}$$

Squaring and rearranging (4.71) yields the Mott and Gurney relationship for the space-charge-limited current resulting from the injection of electrons from the source region:

$$I_{SCL} \cong \frac{9\epsilon_s \mu_n A V_D^2}{8L^3} . \tag{4.72}$$

Similarly, for a gap-type semiconductor structure with p^+ drain and source regions, the space-charge-limited current resulting from the injection of holes from the source region with negative applied drain voltage will be given approximately by

$$I_{SCL} \cong -\frac{9\epsilon_s \mu_p A V_D^2}{8L^3} , \tag{4.73}$$

where μ_p is the mobility of the injected holes in the gap region.

Once again considering the case of an n-channel MOSFET fabricated on a very high-resistivity p-type substrate with a small drain-to-source spacing, it is evident that space-charge-limited current flow from drain to source will be observed when the device is operated with an applied drain voltage sufficiently high to achieve punch-through. Because of the dependence of the space-charge-limited current on the square of the drain voltage, as given by (4.72), one would expect a transition from the saturating pentode-like characteristics of conventional MOSFETs to triode-like characteristics dominated by the space-charge-limited current as the channel length is made increasingly smaller. Such a transition has been predicted theoretically by Geurst,[6] and Neumark and Rittner[7,8] and has been observed experimentally by a number of researchers. This transition can easily be seen in the photographs of Figures 4.14 through 4.16, which show the electrical characteristics of three p-channel MOSFETs that have been fabricated with different drain-to-source spacings on the same nearly intrinsic n-type silicon substrate with all other device parameters held constant. The fact that these devices, as well as similar n-channel configurations, closely follow the V_D^2 dependence as predicted by (4.73) and (4.72), respectively, has been reported by Richman.[9] The square-law dependence is evident at higher current levels where the injected carrier density is much greater than the background concentration of thermally generated carriers in the gap region.

Space-charge-limited current flow will also be observed in MOSFETs with small channel lengths that are fabricated on medium-resistivity substrates when the applied drain voltage is sufficiently high to achieve punch-through. If the applied drain voltage is less than V_{pt}, however, no space-charge-limited current will flow and the observed electrical characteristics will merely exhibit incomplete saturation of the drain current, as described in the previous sections. For MOSFETs with relatively large channel lengths fabricated on low-resistivity substrates, V_{pt} will be very large and, in general, avalanche breakdown of the drain diode will be observed before punch-through can occur.

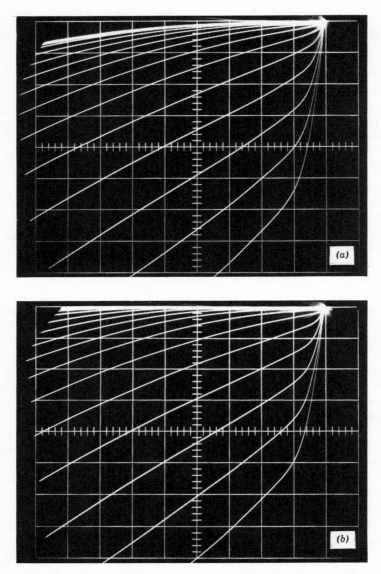

FIGURE 4.14 Electrical characteristics of a *p*-channel MOSFET structure fabricated on a 40,000 Ω-cm *n*-type silicon substrate. $L \cong 0.9$ mil, $W \cong 50$ mils, $T_{ox} \cong 2000$ Å of silicon dioxide; (*a*) vertical scale: drain current: -500 μA/div.; horizontal scale: drain voltage: -5 V/div.; gate voltage: 0 to -15 V, in -1 V steps; $V_S = 0$ V, $T = 25°$C; (*b*) vertical scale: drain current: -500 μA/div.; horizontal scale: drain voltage: -5 V/div.; gate voltage: 0 to -15 V, in -1 V steps; $V_S = +9$ V, $T = 25°$C. (After Richman[9].)

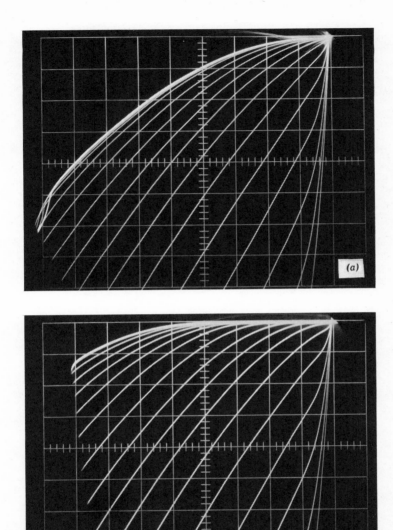

FIGURE 4.15 Electrical characteristics of a p-channel MOSFET structure fabricated on a 40,000 Ω-cm n-type silicon substrate; $L \cong 0.5$ mil, $W \cong 50$ mils, $T_{ox} \cong$ 2000 Å of silicon dioxide; (a) vertical scale: drain current: -500 μA/div.; horizontal scale: drain voltage: -5 V/div.; gate voltage: 0 to -15 V, in -1 V steps; $V_S = 0$ V, $T = 25°$C; (b) vertical scale: drain current: -500 μA/div.; horizontal scale: drain voltage: -5 V/div.; gate voltage: 0 to -15 V, in -1 V steps; $V_S = +9$ V, $T = 25°$C. (After Richman[9].)

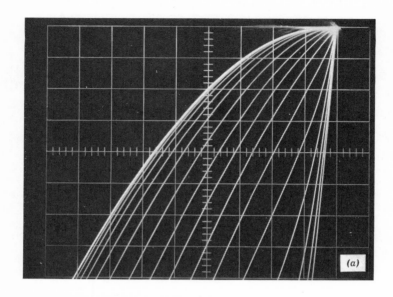

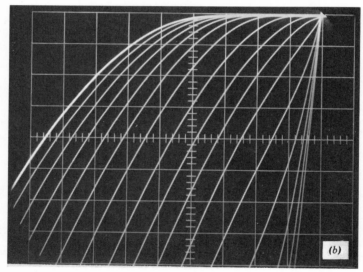

FIGURE 4.16 Electrical characteristics of a p-channel MOSFET structure fabricated on a 40,000 Ω-cm n-type silicon substrate; $L \cong 0.3$ mil, $W \cong 50$ mils, $T_{ox} \cong$ 2000 Å of silicon dioxide; (*a*) vertical scale: drain current: $-500\ \mu\text{A/div.}$; horizontal scale: drain voltage: -2 V/div.; gate voltage: 0 to -15 V, in -1 V steps; $V_S = 0$ V, $T = 25°C$; (*b*) vertical scale: drain current: $-500\ \mu\text{A/div.}$; horizontal scale: drain voltage: -2 V/div.; gate voltage: 0 to -15 V, in -1 V steps; $V_S = +9$ V, $T = 25°C$. (After Richman[9].)

If a MOSFET is operated beyond punch-through, the gate voltage will be able to modulate the flow of space-charge-limited current from drain to source. For an *n*-channel device, the application of a gate voltage which is higher than the threshold voltage V_T will result in the formation of a surface electron inversion layer. Directly below the inversion layer, the depletion region will extend from drain to source. The observed drain-to-source current will consist of two components, the usual drift current flowing through the inversion layer and the (injected) space-charge-limited current flowing through the depletion region. Increasingly negative gate voltage will reduce the drift current component by depleting the surface inversion layer. For gate voltages which are lower than the threshold voltage of the device, the inversion layer will be completely depleted and only space-charge-limited current will flow from drain to source. As the applied gate voltage is made even more negative, it will tend to retard the flow of electrons which have been injected from the source and the observed current will continue to decrease. For large negative values of gate voltage, all surface space-charge-limited current will be eliminated by the effect of the gate field. However, the width of the drain depletion region will be quite large for a high-resistivity substrate, and a substantial amount of space-charge-limited current will flow from drain to source by way of a path through the bulk of the substrate, as illustrated in Figure 4.17. This current will be far enough removed from the silicon surface to remain virtually unaffected by the gate field; consequently, complete cut-off of drain-to-source conduction will not be achieved. This can easily

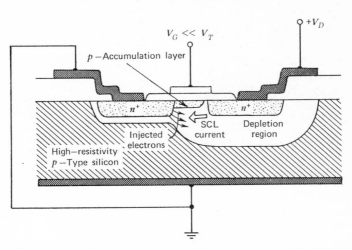

FIGURE 4.17 Bulk space-charge-limited current flow beyond punch-through in an *n*-channel MOSFET operating with zero substrate-to-source potential and a large negative applied gate-to-source voltage.

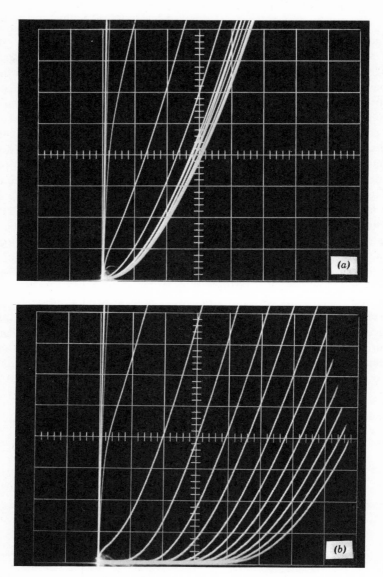

FIGURE 4.18 Electrical characteristics of an *n*-channel MOSFET structure fabricated on a 45,000 Ω-cm *p*-type silicon substrate; $L \cong 0.8$ mil, $W \cong 50$ mils, $T_{ox} \cong 2000$ Å of silicon dioxide; (*a*) vertical scale: drain current: $+100\ \mu$A/div.; horizontal scale: drain voltage: $+5$ V/div.; gate voltage: 0 to -14 V, in -1 V steps; $V_S = 0$ V, $T \doteq 25°$C. (*b*) vertical scale: drain current: $+100\ \mu$A/div.; horizontal scale: drain voltage: $+5$ V/div.; gate voltage: 0 to -14 V in -1 V steps; $V_S = -10$ V, $T = 25°$C. (After Richman[9].)

be seen in the device characteristics of Figures 4.14 through 4.16 with $V_S = 0$ V. The application of a reverse substrate-to-source potential, however, has been observed to retard the space-charge-limited current flowing through the bulk path and to enable complete cut-off of drain-to-source conduction to be achieved over a much wider range of drain voltages.[9] This effect can also be seen in Figures 4.14 through 4.16, upon the application of a $+9$ V substrate-to-source voltage. While the effect of reverse substrate bias upon the drift component of the drain-to-source current is very small for a MOS-FET fabricated on a high-resistivity substrate (see Section 2.3) it can have an appreciable effect on the space-charge-limited current flow beyond punch-through. The charge density contained in the surface depletion region per unit area is so small for a device fabricated on a high-resistivity substrate that even a relatively large increase in its magnitude with the application of reverse substrate-to-source bias will not appreciably affect the threshold voltage of the device. On the other hand, under punch-through conditions, the reverse-biased substrate will act as a grid, retarding the flow of carriers injected from the source region and thereby establishing cut-off over a wide range of applied drain voltages. The major effect that reverse substrate bias can have upon the observed space-charge-limited current is illustrated in Figure 4.18 for an n-channel MOSFET fabricated on a nearly-intrinsic p-type silicon substrate. (It is interesting to note that the characteristics of the device shown in Figure 4.18 with the substrate-to-source potential set equal to -10 V closely resemble those of a vacuum tube triode with an amplification factor approximately equal to 5 and a plate resistance of about $14 \, \text{k}\Omega$.)

As was pointed out previously, one would expect a gradual transition from the saturating pentode-like characteristics associated with conventional MOSFETs to triode-like characteristics dominated by the space-charge-limited current as the channel length of a device fabricated on a high-resistivity substrate is made increasingly smaller. However, Geurst,[6] using a simplified mathematical model of an insulated-gate field-effect transistor structure, has shown that this transition will be strongly dependent upon geometrical considerations. In particular, both Geurst[6] and Neumark and Rittner[7,8] have shown theoretically that the ratio of the thickness of the gate insulator to the actual spacing between the drain and source electrodes plays a very important role in determining the observed output conductance of the device and the extent to which the space-charge-limited current will dominate the actual current-voltage characteristics. By extending the basic Geurst theory, Neumark and Rittner showed that as the ratio of the drain-to-source spacing to the thickness of the gate insulator decreases, the transition to triode-like characteristics will be observed at lower values of drain voltage and the output conductance will *increase*. Experimental verification of these predictions can be seen in Figure 4.19, which shows the current-voltage

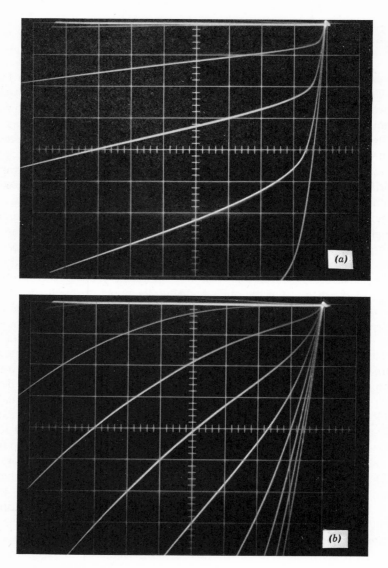

FIGURE 4.19 (*a*) Current-voltage characteristics associated with a *p*-channel MOSFET fabricated on a 7.5 Ω-cm *n*-type silicon substrate with an actual drain-to-source spacing of approximately 0.25 mil and a gate insulator consisting of approximately 600 Å of silicon dioxide covered by 400 Å of silicon nitride; vertical scale: drain current: -100 μA/div.; horizontal scale: drain voltage: -2 V/div.; gate voltage: 0 to -4 V, in $-\frac{1}{2}$ V steps; $V_S = 0$ V, $T = 25°$C; (*b*) Current-voltage characteristics associated with a *p*-channel MOSFET simultaneously fabricated on the same 7.5 Ω-cm *n*-type silicon substrate with virtually the same values of channel length and channel width, but with a gate insulator consisting approximately of 15,000 Å of silicon dioxide covered by 400 Å of silicon nitride; vertical scale: drain current: -100 μA/div.; horizontal scale: drain voltage: -2 V/div.; gate voltage: 0 to -60 V, in -4 V steps; $V_S = 0$ V, $T = 25°$C.

characteristics associated with two *p*-channel MOSFETS which have been simultaneously fabricated on a single *n*-type silicon substrate with virtually the same values of channel length and channel width, but with greatly differing thicknesses of gate insulator. It can be seen that the device fabricated with the relatively thin gate insulator exhibits saturating pentode-like characteristics, while the device fabricated with the much thicker gate insulator exhibits triode-like characteristics over the same range of drain current and drain voltage.

4.4 CONDUCTION BEYOND SATURATION: AVALANCHE BREAKDOWN OF THE DRAIN DIODE AT LARGE VALUES OF APPLIED DRAIN VOLTAGE

If the voltage applied to the drain electrode of a MOS device which is operating in the region of saturated drain current is steadily increased, the

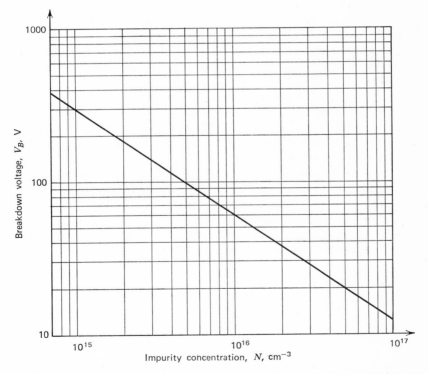

FIGURE 4.20 Avalanche breakdown voltages versus impurity doping concentration in the lightly doped side of the junction for one-dimensional silicon step-junctions (after McKay[10]).

drain voltage will eventually become high enough to cause avalanche break-down of the drain diode, and the drain current will be observed to increase rapidly. It is well-known that the voltage at which avalanche breakdown will occur in a reverse-biased *p-n* junction will be strongly dependent upon the doping concentration in the high-resistivity side of the junction and, in particular, will decrease as the doping concentration is increased. This can easily be seen in Figure 4.20, which summarizes experimental data reported by McKay[10] for avalanche breakdown voltages associated with n^+p or p^+n one-dimensional silicon step-junction diodes as a function of the effective impurity concentration in the lightly doped side of the junction. However, the values of drain voltage at which avalanche breakdown is typically observed in MOSFET structures are usually considerably lower than one might expect on the basis of the data shown in Figure 4.20.

In general, besides being strongly dependent on the impurity doping concentration in the substrate, the breakdown voltage associated with the drain diode of typical MOSFET devices has also been found to be greatly influenced by the points of maximum curvature (hence the junction depth) of the drain diffusion and, also, by the proximity of the gate electrode to the depletion region located near the drain end of the channel. The latter two effects are primarily responsible for the lowering of the breakdown voltage of the drain diode below the values predicted by Figure 4.20.

4.4.1 Avalanche Breakdown in One-Dimensional Abrupt *p-n* Junctions

For all practical purposes, when a *p-n* junction is reverse-biased, virtually all of the applied voltage is impressed across the depletion region; conse-quently, all of the resulting electric field is confined to this region. In the case of a one-dimensional step-junction, the electric field intensity has been shown to vary linearly with distance within the depletion region, reaching a maxi-mum value at the junction boundary itself.[11] As the voltage impressed across the reverse-biased junction is increased, the width of the depletion region increases with the square root of the total voltage across the junction and, as a result, the electric field intensity within the depletion region also increases. When the reverse bias is sufficiently high to produce a critical electric field intensity within the depletion region, avalanche breakdown will occur. (The magnitude of the critical field intensity at breakdown will vary slowly with the doping concentration in the lightly doped side of the junction. Typically, for a doping concentration of $N \cong 10^{16}/cm^3$ in the lightly doped side, for a one-dimensional silicon step junction, the maximum field at breakdown will be approximately 4×10^5 V/cm).[12] The avalanche breakdown process is a result of the extremely high electric field within the depletion region acting to rapidly accelerate any free carriers that happen to be present. On the average, each carrier will travel a distance λ, the mean free path, while it is being accelerated by the electric field. Then it will collide with either a silicon atom in the lattice or with an impurity atom and will be scattered. If the electric field intensity is sufficiently high (i.e., equal to or greater than the critical field), the carrier will have been accelerated to such a high velocity that additional carriers may be generated when the next collision takes place. For example, an electron moving through the depletion region may be accelerated by the high field to a velocity (and a corresponding kinetic energy) sufficiently great to strip an outer-orbit electron from the next atom it undergoes a collision process with. Now, there will be two electrons, both of which will be accelerated by the high electric field to collide with other atoms and, once again, strip off additional electrons. The process will continue and a chain-reaction will take place. The resulting rapid increase in the free electron concentration will give rise to a sharp increase in the observed reverse leakage current associated with the junction. This rapid increase in the reverse current above a certain value of reverse bias is known as *avalanche breakdown*. The avalanche process will occur only for electric fields of sufficient strength to accelerate an electron to a velocity v_o as it travels a distance equal to the mean free path, such that just before the collision takes place

$$\tfrac{1}{2}m_e(v_o{}^2 - v_f{}^2) \geqq E_I, \tag{4.74}$$

where m_e is the mass of an electron, v_f is the final velocity of the first electron just after the collision process, and E_I is the ionization energy of the bound outer-orbit electron of the silicon atom with which the electron has collided. The higher the reverse bias and, consequently, the higher the electric field intensity within the depletion region, the more readily this avalanche process occurs and the more reverse current flows. When the reverse bias reaches the breakdown voltage V_B, complete breakdown occurs. For an asymmetric step junction in which the doping concentration on one side of the junction is orders of magnitude greater than the doping concentration on the other side, the depletion region, for all practical purposes, will spread almost completely into the lightly doped side with increasing reverse bias. Thus it follows that the width of the depletion region, the electric field intensity within the depletion region, and the resulting value of the avalanche breakdown voltage will, to first order, be dependent only on the doping concentration in the lightly doped side of the junction and will be almost completely independent of the concentration in the heavily doped side.

4.4.2   The Effect of the Curvature of a *p-n* Junction on the Breakdown Voltage

As a result of geometric factors, diffused planar *p-n* junctions will typically exhibit avalanche breakdown voltages which are considerably lower than the values predicted by the data shown in Figure 4.20 for one-dimensional silicon step-junctions. When a *p-n* junction is formed by conventional planar processing techniques by diffusing a high concentration of impurity atoms through a rectangular "window" in a masking layer of silicon dioxide and into a silicon substrate of opposite polarity, the impurities will tend to diffuse laterally, from the edges of the "window," a distance which is approximately

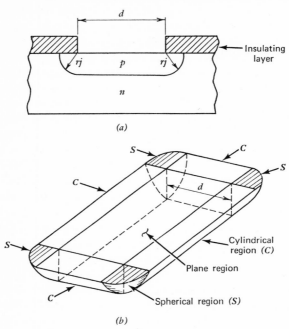

FIGURE 4.21 Lateral diffusion effects during the diffusion of impurity atoms into a lightly doped substrate through a rectangular window in the masking oxide layer. (*a*) cross-sectional view; r_j is the radius of curvature at the periphery; (*b*) the formation of approximately cylindrical and spherical regions by diffusing through a rectangular mask. (After Simon Sze, *Physics of Semiconductor Devices*, John Wiley and Sons, New York, p. 85.)

equal to the distance they will diffuse downward into the bulk of the substrate. Consequently, the amount of lateral diffusion in such a planar structure will be comparable to the junction depth and, as can be seen in Figure 4.21, diffusion through a rectangular window will actually result in a structure in which the junction itself is a combination of flat, cylindrical, and spherical regions.[13] It can be seen in Figure 4.21 that the radius of curvature r_j in the cylindrical and spherical regions will be approximately equal to the junction depth and if the junction is extremely shallow the radius of curvature will be quite small.

It is well-known that, for a given doping concentration in the lightly doped side, the avalanche breakdown voltage associated with a diffused planar junction will *decrease* with decreasing junction depth. This is a direct result of the fact that, as has been previously discussed, the avalanche breakdown voltage will be a very strong function of the electric field intensity present within the depletion region of the junction. If the field distribution in the depletion region is uneven, the point where the electric field strength is highest will break down first. Certain junction geometries will alter the field distribution in the depletion region and will concentrate or *intensify* the electric field in particular areas, making these areas especially susceptible to early breakdown.[14–17] This effect is illustrated in Figure 4.22. The one-dimensional step-junction shown in Figure 4.22a is characterized, under reverse-bias conditions, by a depletion region of uniform width and an electric field distribution within the depletion region that does not vary with position (in the lateral direction) over the surface of the junction. However, as can be seen in Figure 4.22b, in a shallow diffused planar junction, the electric field strength is highest near the regions of sharpest curvature. Because of the higher electric field strength for a given value of reverse bias, these regions will experience avalanche breakdown at lower voltages. The smaller the radius of curvature in a particular region of the junction, the greater the electric field intensification will be in that region, and the lower the breakdown voltage associated with the junction. A good analogy is the case of a perfect conductor in the presence of a strong incident electric field resulting from the application of a relatively high voltage to a nearby electrode. Breakdown (a spark across the gap) will occur at the *points* on the conductor, or the regions of sharpest curvature. Analogously, avalanche breakdown will occur at the regions of sharpest curvature associated with the *p-n* junction. In general, the observed value of avalanche breakdown voltage for a shallow diffused *p-n* junction will be substantially less than that predicted for a similar one-dimensional diode structure if the smallest radius of curvature within the structure is comparable or less than the width of the depletion region in the one-dimensional structure.

Sze and Gibbons[17] have calculated the effect on the avalanche breakdown voltage of junction curvature for both cylindrical and spherical abrupt silicon step-junctions at room temperature. The results of these calculations are shown in Figure 4.23 as a function of the impurity doping concentration in the high-resistivity side of the junction. For the case of an infinitely large radius of curvature, corresponding to the one-dimensional approximation, it can be seen that the calculated values of breakdown voltage approach the data given in Figure 4.20.

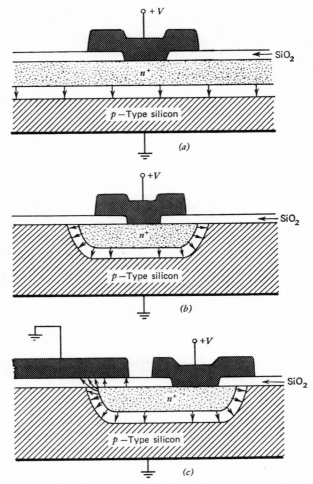

FIGURE 4.22 - (*a*) One-dimensional abrupt silicon *p-n* junction under reverse-bias conditions with uniform electric field distribution over the surface of the entire junction; (*b*) electric field intensification at regions of greatest curvature in an abrupt diffused planar silicon *p-n* junction; (*c*) electric field intensification within the depletion region in an abrupt diffused planar silicon *p-n* junction resulting from the close proximity of a (grounded) conducting electrode.

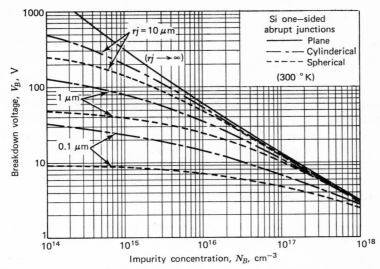

FIGURE 4.23   Avalanche breakdown voltage versus impurity doping concentration in the high resistivity side of abrupt silicon *p-n* junctions for both cylindrical and spherical structures. r_j is the radius of curvature. (After Sze and Gibbons[17].)

4.4.3   The Effect of the Proximity of a Conducting Electrode on the Avalanche Breakdown Characteristics of a Diffused *p-n* Junction

Perhaps the most important consideration in the determination of the avalanche breakdown voltage for the drain region of a MOSFET is the proximity of the gate electrode to the depletion region which exists between the drain diffusion and the channel region for drain voltages beyond pinch-off. Typically, the observed drain avalanche breakdown voltage for conventional MOSFET structures with diffused junction depths in the 1 to 3 μ range is considerably below what would be expected on the basis of considering only the effect of junction curvature and, in particular, the data shown in Figure 4.23. This further reduction in the avalanche breakdown voltage associated with the drain diode in a MOSFET has been observed to be a strong function of the thickness of the gate insulator and has been shown to be a direct function of the separation distance between the edge of the drain diffusion and the overlying gate electrode. (For MOSFETs fabricated with relatively thick gate insulators, good correlation is obtained between the data shown in Figure 4.23 and the actual drain breakdown voltage.)

As can be seen in Figure 4.22c, the presence of the gate electrode redistributes the electric field lines near the surface of the drain depletion region, creating a region of extremely high field strength and correspondingly lower breakdown voltage.[18] This reduction in the breakdown voltage is observed if the gate electrode is grounded or even if it is left floating. For thin silicon dioxide gate insulators, the effect of varying the substrate resistivity over a wide range of doping concentrations on the breakdown voltage has been shown to be almost negligible.

As a result of the above effect, drain breakdown voltages associated with conventional MOSFETs are typically on the order of 25 to 40 V. This early breakdown voltage has been used extensively to provide "Zener protection" for the thin gate insulators of MOS devices from damage resulting from high-voltage "spikes" and static charge rupture and, as opposed to a conventional diffused zener structure, enables one to provide effective protection at relatively low voltages even on high-resistivity substrates. The protection is achieved merely by placing structures similar to the one shown in Figure 4.22c, with grounded gate electrodes, in parallel with all inputs leading directly into gate electrodes. For all input voltages greater than the avalanche breakdown voltage, the diode will conduct and limit the maximum voltage which is applied to the gate electrode. For all input voltages below the avalanche breakdown voltage, the only degrading effect will be a slight lowering in the input impedance associated with the gate structure. Of course, another restriction will be that an opposite polarity gate voltage can no longer be applied to the MOSFET because the diode that has been placed in parallel with the gate will become forward-biased and will begin to conduct. However, in digital integrated MOS circuits where the use of enhancement-type devices eliminates the need for opposite-polarity voltages to achieve "on-off" operation, the restriction is relatively unimportant.

The reason that the proximity of the gate electrode to the drain depletion region has such a pronounced effect on the observed breakdown characteristics of the drain diode can best be understood by considering the boundary conditions to Maxwell's equations for electrostatic fields in differential form.[19] First, the tangential component of the electric field intensity must be continuous across an interface between two regions. That is,

$$\mathbf{n} \, x \, (\mathscr{E}_{t1} - \mathscr{E}_{t2}) = 0, \tag{4.75}$$

where $\mathbf{n}$ is the unit vector normal to the interface between the two regions and $\mathscr{E}_{t1}$ and $\mathscr{E}_{t2}$ are the tangential components of the electric field intensity in the first and second regions, respectively. Next, the boundary condition on the normal component of the electric field intensity across the interface between the two regions is

$$\mathbf{n} \cdot (\epsilon_1 \mathscr{E}_{n1} - \epsilon_2 \mathscr{E}_{n2}) = \sigma, \tag{4.76}$$

where ϵ_1 and ϵ_2 are the dielectric constants associated with the first and second regions, $\mathscr{E}_{n1}$ and $\mathscr{E}_{n2}$ are the normal components of the electric field intensity in these regions, and σ is the total charge density per unit area located at the interface. Finally, the *constituent-relation* for a conductor which obeys Ohm's law is

$$\mathbf{J} = \sigma_C \mathscr{E}, \tag{4.77}$$

where J is the current density per unit area, σ_C the conductivity of the material through which the current is flowing, and $\mathscr{E}$ the electric field within the material. Since the electric field intensity in a perfectly conducting material ($\sigma_C \to \infty$) must approach zero to maintain a finite current flow, it follows that if the gate electrode consists of a highly conductive material and is assumed to be the *second* region while the underlying gate insulator is assumed to be the *first* region, then

$$\mathscr{E}_{n2} = \mathscr{E}_{t2} = \mathscr{E}_{t1} = 0 \tag{4.78}$$

or, in other words, *the electric field intensity in the gate insulator must, in the ideal case of a perfectly conducting gate region, be perpendicularly incident on the gate electrode.*

Referring once again to Figure 4.22*b*, it can be easily seen that, in the absence of the gate electrode, one would expect the electric field lines within the depletion region near the surface of the silicon to be approximately directed *parallel to the surface*. However, one can conclude from the above argument that, by placing the gate electrode in close proximity to the surface portion of the depletion region, the electromagnetic requirement that the electric field lines in the gate insulator be *perpendicular* to the surface near the gate electrode (as shown in Figure 4.22*c*) results in severe distortion and intensification of the electric field lines in the depletion region directly below the gate electrode and a corresponding reduction in the avalanche breakdown voltage of the device. As the thickness of the gate insulator is increased, the amount of distortion and field intensification decreases and, in the limit of an infinitely thick gate insulator separating the gate electrode from the drain depletion region, the avalanche breakdown voltage of the drain diode approaches the value expected solely on the basis of junction curvature considerations.

4.4.4 The Effect of an Applied Gate Voltage on the Breakdown Voltage of the Drain Diode

Since the presence of a highly conductive electrode in close proximity to the drain junction has such a pronounced effect on the avalanche breakdown characteristics of the junction, it follows that the breakdown characteristics will also be strongly affected by variations in the voltage which is applied to the gate electrode.[18] For example, the application of a negative voltage to the gate electrode of an *n*-channel MOSFET fabricated with a relatively thin gate insulator which is operating in the avalanche region will tend to further increase the potential difference existing between the gate and the drain electrodes. This will result in an even greater increase in the electric field intensity in the depletion region near the silicon surface directly below the gate electrode. Consequently, the avalanche mechanism in this region will be aided by the increased field and the breakdown voltage will be observed to decrease with increasingly negative applied gate voltage. Conversely, if a small positive voltage is applied to the gate electrode of the same n-channel MOSFET, the field strength in the critical high-field region will be reduced and the avalanche breakdown voltage will be observed to increase. Although MOSFETs will only very rarely be intentionally biased into the avalanche region, other MOS-type structures utilizing the gate-voltage-dependent avalanche breakdown voltage mechanism described above have been studied by Shockley and Hooper,[20] Nathanson,[21] and others. Electric field modulation of the avalanche breakdown voltage and actual amplification have been achieved.

Once again considering the case of an n-channel MOSFET, it has been shown that as the applied gate-to-source potential is made increasingly more negative, the avalanche breakdown voltage associated with the drain junction is observed to decrease. However, if the doping concentration at the surface of the low-resistivity drain diffused region is sufficiently low (on the order of 10^{19} to $10^{18}/cm^3$), another mechanism begins to contribute to the breakdown process for relatively large values of negative applied gate voltage. If the gate electrode considerably overlaps the drain diffused region, the electric field in the thin gate insulator separating these two regions can become sufficiently large to create a p-type surface inversion layer within that part of the drain diffusion which is overlapped by the gate. (The value of gate voltage at which the inversion layer will form, just as in the case of the threshold voltage associated with the channel region, will be a function of the doping concentration at the surface of the diffused drain region, the fixed positive interface charge density, the thickness of the insulator which separates the drain from the gate, and the work function difference which exists between the diffused drain region and the gate electrode. If the doping concentration in the diffused region is too high, the voltage required for the inversion layer to form will be so high that the insulating layer will rupture before surface inversion can take place.)

As is illustrated in Figure 4.24, the application of a highly negative gate voltage can form a p^+ inversion layer at the surface of an n^+ diffused region directly under the gate electrode. Similarly, as shown in Figure 4.25, an n^+ surface inversion layer can be formed in a p^+ diffused region, directly under an overlying gate electrode, for very large positive gate voltages. In both structures, a field-induced p^+-n^+ junction is formed which can undergo

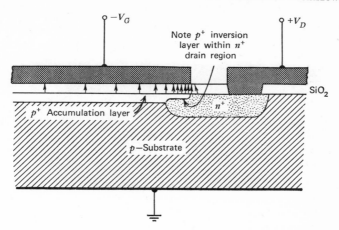

FIGURE 4.24 Condition for tunnel emission breakdown resulting from the action of the gate field in an n^+p MOS structure. (After Richman[18].)

breakdown at a relatively low value of reverse bias by means of tunnel emission.[18]

The depletion region in the reverse-biased field-induced p^+-n^+ junction will be extremely thin. If both the doping concentration in the heavily diffused region and the electric field intensity in the gate insulator are sufficiently high, both sides of the junction will be *degenerate*. (That is, the Fermi level will lie *above* the conduction band edge in the n^+ side of the junction and *below* the valence band edge in the p^+ side.) With the applied drain-to-substrate voltage set equal to zero, no current will flow in either of the two structures, and a state of equilibrium will exist. The energy band diagram associated with the p^+-n^+ junctions formed in Figures 4.24 and 4.25 with zero drain-to-substrate voltage is illustrated in Figure 4.26. When the field-induced p^+-n^+ junction of either structure is reverse-biased through the application of a drain-to-substrate voltage of the appropriate polarity, many electrons in the valence band of the p^+ region will have *higher* energies than unoccupied electronic states in the conduction band of the n^+ region, as is illustrated in Figure 4.27. Although the depletion region presents a potential barrier that prohibits them from occupying these states, it is well known that, for sufficiently thin potential barriers, quantum-mechanical considerations predict that it will be possible for a number of the electrons in the higher-energy states of the valence band to tunnel through the thin depletion region into the empty states in the n^+ side of the junction. This results in a flow of current that will increase with increasing reverse bias and tunnel emission breakdown will occur.

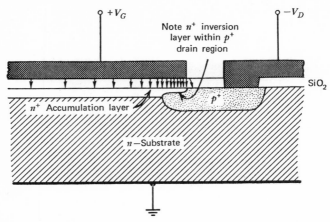

FIGURE 4.25   Condition for tunnel emission breakdown resulting from the action of the gate field in a p^+n MOS structure. (After Richman[18].)

It should be noted that tunnel emission breakdown is only rarely observed in typical MOS structures and occurs only for a rather narrow range of doping concentrations at the surface of the diffused drain region. If the doping concentration is *too high*, the gate voltage required to achieve surface inversion in the drain region will be higher than the breakdown strength of the insulating region and the latter will rupture before inversion takes place.

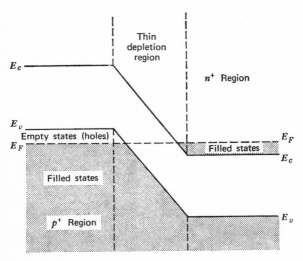

FIGURE 4.26   The p^+-n^+ junction with zero applied voltage. (After Richman[18].)

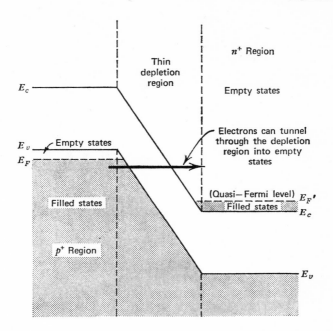

FIGURE 4.27 The p^+-n^+ junction under reverse bias conditions where electron tunneling can take place. (After Richman[18].)

On the other hand, if the doping concentration is *too low*, the drain region will not be degenerate and the previously discussed conditions for electron tunneling will not be achieved.

REFERENCES

1. W. Shockley, A Unipolar Field-Effect Transistor, *Proceedings of the IRE*, Vol. 40, November 1952, pp. 1365–1376.
2. A. S. Grove, O. Leistiko, and W. W. Hooper, Effect of Surface Fields on the Breakdown Voltage of Planar Silicon *p-n* Junctions, *IEEE Transactions on Electron Devices*, Vol. ED-14, No. 3, 1967, pp. 157–162.
3. S. R. Hofstein and G. Warfield, Carrier Mobility and Current Saturation in the MOS Transistor, *IEEE Transactions on Electron Devices*, Vol. ED-12, No. 3, 1965, pp. 129–138.
4. S. R. Hofstein and F. P. Heiman, The Silicon Insulated-Gate Field-Effect Transistor. *Proceedings of the IEEE*, Vol. 51, No. 9, September 1963, pp. 1190–1202.
5. N. F. Mott and R. W. Gurney, *Electronic Processes in Ionic Crystals*, second edition, Clarendon Press, Oxford, 1948, p. 172.

6. J. A. Geurst, Theory of Insulated-Gate Field-Effect Transistors Near and Beyond Pinch-Off, *Solid State Electronics*, Vol. 9, 1966, pp. 129–142.

7. E. S. Rittner and G. F. Neumark, Theory of the Surface Gate Dielectric Triode, *Solid State Electronics*, Vol. 9, 1966, pp. 895–898.

8. G. F. Neumark and E. S. Rittner, Transition from Pentode- to Triode-Like Characteristics in Field-Effect Transistors, *Solid State Electronics*, Vol. 10, 1967, pp. 299–304.

9. P. Richman, Modulation of Space-Charge-Limited Current Flow in Insulated-Gate Field-Effect Tetrodes, *IEEE Transactions on Electron Devices*, Vol. ED-16, No. 9, 1969, pp. 759–766.

10. K. G. McKay, Avalanche Breakdown in Silicon, *Physical Review*, Vol. 94, May 15, 1954, pp. 887–884.

11. A. B. Phillips, *Transistor Engineering*, McGraw-Hill Book Co., New York, 1962, pp. 108–112.

12. S. M. Sze and G. Gibbons, Avalanche Breakdown Voltages of Abrupt and Linearly Graded *p-n* Junctions in Ge, Si, GaAs, and GaP, *Applied Physics Letters*, Vol. 8, 1966, p. 111.

13. T. P. Lee and S. M. Sze, Depletion Layer Capacitance of Cylindrical and Spherical *p-n* Junctions, *Solid State Electronics*, Vol. 10, 1967, p. 1105.

14. H. L. Armstrong, A Theory of Voltage Breakdown of Cylindrical *p-n* Junctions with Applications, *IRE Transactions on Electron Devices*, January 1957, pp. 15–16.

15. H. L. Armstrong, E. D. Metz, and I. Weiman, Design Theory and Experiments for Abrupt Hemispherical *p-n* Junction Diodes, *IRE Transactions on Electron Devices*, April 1956, pp. 86–92.

16. F. P. Heiman and S. R. Hofstein, Metal Oxide Semiconductor Field Effect Transistors, *Electronics*, November 1964, pp. 50–61.

17. S. M. Sze and G. Gibbons, Effect of Junction Curvature on Breakdown Voltages in Semiconductors, *Solid State Electronics*, Vol. 9, 1966, p. 831.

18. P. Richman, *Characteristics and Operation of MOS Field-Effect Devices*, McGraw-Hill Book Co., New York, 1967, pp. 63–73.

19. R. M. Fano, L. J. Chu, and R. B. Adler, *Electromagnetic Fields, Energy and Forces*, John Wiley and Sons, New York, 1960, pp. 86–89 and 184–188.

20. W. Shockley and W. W. Hooper, The Surface Controlled Avalanche Transistor, 1964 Western Electronic Show and Convention, Los Angeles, California, August 25–28, 1964.

21. H C. Nathanson, A High Field Triode, *Solid State Electronics*, Vol. 8, 1965, pp. 349–363.

BIBLIOGRAPHY

Armstrong, G. A., and J. A. Magowan, The Distribution of Mobile Carriers in the Pinch-Off Region of an Insulated-Gate Field-Effect Transistor and its Influence on Device Breakdown, *Solid State Electronics*, Vol. 14, 1971, pp. 723–733.

Armstrong, G. A., and J. A. Magowan, Pinch-Off in Insulated-Gate Field-Effect Transistors, *Solid State Electronics*, Vol. 14, 1971, pp. 760–763.

Asakawa, T., and N. Tsubouchi, Second Breakdown in MOS Transistors, *IEEE Transactions on Electron Devices*, Vol ED-13, No. 11, November 1966, pp. 811–812.

Bloodworth, G., Four-Terminal Operation of MOS Transistors, *Proceedings of the IEE* (Great Britain), Vol. 113, No. 10, October 1966, pp. 1587–1595.

Chiu, T. L., and C. T. Sah, Correlation of Experiments with a Two-Section-Model Theory of the Saturation Drain Conductance of MOS Transistors, *Solid State Electronics*, Vol. 11, 1968, pp. 1149–1163.

Das, M. B., Charge-Control Analysis of MOS and Junction-Gate Field-Effect Transistors, *Proceedings of the IEE* (Great Britain), Vol. 113, No. 10, October 1966, pp. 1565–1570.

Denda, S., and M. A. Nicolet, Pure Space-Charge-Limited Electron Current in Silicon, *Journal of Applied Physics*, Vol. 37, No. 6, May 1966, pp. 2412–2424.

Frohman-Bentchkowsky, D., and A. S. Grove, Conductance of MOS Transistors in Saturation, *IEEE Transactions on Electron Devices*, Vol. ED-16, No. 1, January 1969, pp. 108–113.

Greene, R., and T. Soldano, Increasing the Accuracy of MOS Calculations, *Proceedings of the IEEE*, Vol. 53, No. 9, September 1965, pp. 1241–1242.

Grove, A. S., and D. J. Fitzgerald, The Origin of Channel Currents Associated with p+ regions in Silicon, *IEEE Transactions on Electron Devices*, Vol. ED-12, No. 12, December 1965, pp. 619–626.

Hagenlocher, A. K., Space-Charge-Limited Currents in High-Resistivity *p*-type Silicon, *Applied Physics Letters*, Vol. 10, No. 4, 1967, pp. 119–121.

Hofstein, S. R., and G. Warfield, The Insulated-Gate Tunnel Junction Triode, *IEEE Transactions on Electron Devices*, Vol. ED-12, No. 2, February 1965, pp. 66–76.

Ihantola, H. K. J., and J. L. Moll, Design Theory of a Surface Field-Effect Transistor, *Solid State Electronics*, Vol. 7, 1964, pp. 423–430.

Pao, H. C., and C. T. Sah, Effects of Diffusion Current on Characteristics of Metal-Oxide (Insulator)-Semiconductor Transistors, *Solid State Electronics*, Vol. 9, 1966, pp. 927–937.

Popa, A., An Injection Level Dependent Theory of the MOS Transistor in Saturation, *IEEE Transactions on Electron Devices*, Vol. ED-19, No. 6, June 1972, pp. 774–781.

Reddi, V. G. K., and C. T. Sah, Source to Drain Resistance Beyond Pinch-Off in Metal-Oxide-Semiconductor Transistors (MOST), *IEEE Transactions on Electron Devices*, Vol. ED-12, No. 3, 1965, pp. 139–141.

Smith, D. P., and J. G. Linvill, An Accurate Short-Channel IGFET Model for Computer-Aided Circuit Design, presented at the 1971 International Solid State Circuits Conference, Philadelphia, Pennsylvania, February 1971.

Vandorpe, D., J. Borel, G. Merckel, and P. Saintot, An Accurate Two-Dimensional Numerical Analysis of the MOS Transistor, *Solid State Electronics*, Vol. 15, 1972, pp. 547–557.

Weinerth, H., Silicon Diode Breakdown in the Transition Range Between Avalanche Effect and Field Emission, *Solid State Electronics*, Vol. 10, 1967, pp. 1053–1062.

Wright, G. T., Space-Charge-Limited Insulated-Gate Surface-Channel Transistor, *Electronics Letters*, Vol. 4, No. 21, October, 1968, pp. 462–464.

Wright, G. T., Theory of the Space-Charge-Limited Surface-Channel Dielectric Triode, *Solid State Electronics*, Vol. 7, 1964, pp. 167–175.

Zuleeg, R., and P. Knoll, Space-Charge-Limited Currents in Heteroepitaxial Films of Silicon Grown on Sapphire, *Applied Physics Letters*, Vol. 11, No. 6, 1967, pp. 183–185.

Zuleeg, R., and P. Knoll, A Thin-Film Space-Charge-Limited Triode, *Proceedings of the IEEE*, Vol. 54, No. 9, September 1966, pp. 1197–1198.

PROBLEMS

4.1 Show that when $\Phi \ll 1$ and $V_G - V_{FB} - 2\phi_F \gg 1$, (4.40) can be approximated by (4.41).

4.2 Consider an n-channel MOSFET fabricated on a p-type silicon substrate with $N_A = 10^{15}/\text{cm}^3$. The thickness of the silicon dioxide gate insulator is 1000 Å. With the applied gate voltage such that $V_G - V_{FB} = +15$ V, calculate the drain voltage at which saturation of the drain current will be observed. Now, consider a similar device with $N_A = 10^{15}/\text{cm}^3$, $V_G - V_{FB} = +15$ V, and $T_{ox} = 2000$ Å of silicon dioxide. Once again, calculate the drain voltage at which saturation of the drain current will be observed. Comment on your results.

4.3 Assume the spreading of the depletion region that forms at the drain-channel junction beyond pinch-off is given by

$$\Delta L \cong \left[\frac{12\epsilon_s(V_D - V_{D\text{sat}})}{qa} \right]^{1/3},$$

that is, as for a linearly graded junction for which the effective impurity concentration varies linearly with y near the drain end of the channel and is proportional to a constant a, the grade constant. Derive an expression for $I_{D\text{sat}'}$, the drain current under conditions of incomplete current saturation, considering channel length modulation only. (*Note*: in a linearly graded junction, unlike in a step junction, the depletion region spreads equally in both directions with increasing reverse bias.)

4.4 Consider an n-channel MOSFET fabricated on a moderate-resistivity p-type silicon substrate with a drain-to-source spacing L comparable to the maximum depth of the surface depletion region, $x_{d_{\max}}$. Assuming that the drain field which couples through the depletion region and terminates in the channel can

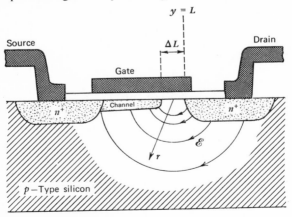

FIGURE 4.28

be represented by semicircles centered at $y = L - \frac{1}{2}\Delta L$, as shown in Figure 4.28, show that the total drain-to-channel coupling capacitance will be approximately equal to

$$C_{dct} \cong \frac{\epsilon_s W}{\pi} \ln\left(\frac{2L}{\Delta L}\right).$$

If L is equal to 10 μ and, initially, ΔL is equal to 1 μ, find the fractional change in C_{dct} when the drain voltage is increased so that ΔL equals 2 μ. Thus demonstrate that C_{dct} is a slowly varying function of the applied drain voltage.

4.5 Describe qualitatively, for a MOSFET operating beyond punch-through, the behavior as a function of temperature of: (a) the drain-to-source conductance, (b) the current levels at which the onset of triode-like operation (with I_D proportional to $V_D{}^2$) will be observed, and (c) the magnitude of the bulk space-charge-limited current that cannot be cut-off by the gate field with the substrate-to-source potential set equal to zero.

4.6 When the effects of the quantities Q_{SS} and $\phi_{MS'}$ upon the spreading of the surface depletion region from the drain end of the channel are taken into account, will the punch-through voltage V_{pt} be greater than or less than the value predicted by first order theory (see eq. 4.62) for an n-channel MOSFET with $V_G = 0$? Explain your answer. What will be the effect on a p-channel MOSFET with $V_G = 0$?

4.7 Plot the (first-order) punch-through voltage V_{pt}, as given by (4.62), versus the acceptor doping concentration in the p-type substrate of an n-channel MOSFET for the following values of actual drain-to-source spacing: $L = 0.2$, 0.3, 0.4, 0.6, 0.8 and 1.0 mil.

4.8 Discuss possible fabrication techniques for the construction of a MOSFET structure on a moderately high-resistivity silicon substrate with a drain breakdown voltage approaching that which is predicted by first-order theory. How could the transconductance associated with this structure be kept comparable to a similar device with a relatively low drain breakdown voltage that has been fabricated in a comparable area?

5

The Effect of Temperature Variations on the Electrical Characteristics of MOS Field-Effect Transistors

The two parameters that are commonly used to characterize the current-versus-voltage characteristics of MOSFETs are the quantities β and V_T. In particular, to first-order, the drain current for small values of applied drain voltage has been shown to be approximately given by

$$I_D \cong \beta V_D (V_G - V_T) \tag{4.20}$$

for an n-channel MOSFET, and the saturated drain current has been shown to be given approximately by

$$I_{D\text{sat}} \cong \frac{\beta}{2} (V_G - V_T)^2. \tag{4.41}$$

Note that both (4.20) and (4.41) depend *only* on β, V_T, and the applied voltages. Consequently, it follows that any variation in the electrical characteristics of a MOSFET as a function of temperature can be directly attributed to the variations with temperature of these two quantities.

The β is a direct measure of the gain of the transistor and is defined by

$$\beta \equiv \frac{\epsilon_{ox}\mu W}{T_{ox}L}. \tag{4.12}$$

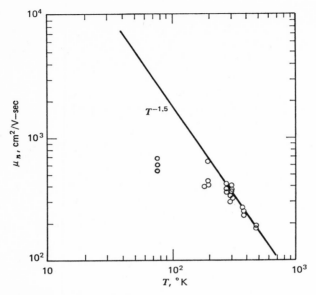

FIGURE 5.1 Effective inversion layer electron mobility as a function of temperature for silicon with $(-Q_{\text{total}}/q) < 10^{12}/\text{cm}^2$ (After Leistiko, Grove, and Sah[1].)

It is apparent that any variations with temperature of the parameter β can be directly attributable to variations in μ, the effective mobility of the carriers in the channel. As the operating temperature of the device is increased, the effective carrier mobility in the channel will *decrease*. Therefore, the quantity β will also be observed to decrease with increasing temperature. The variation of effective carrier mobility in the inversion layer with temperature has been studied experimentally by Leistiko, Grove, and Sah for both *n*- and *p*-channel silicon MOSFETs.[1] As shown in Figures 5.1 and 5.2, the experimental results indicate that for high temperatures the effective mobilities exhibit a $T^{-3/2}$ dependence. However, at lower temperatures of more practical interest, the decrease in carrier mobility with increasing temperature is not as rapid. In the intermediate range of -55 to $+125°C$, the variation of the mobility is better approximated by a T^{-1} dependence.[2] The gain of a MOSFET in this range, as measured by the quantity β, will behave in a corresponding manner.

The threshold voltage, defined as the value of the applied gate-to-source voltage that is required to strongly invert the semiconductor surface, is given approximately by

$$V_T \cong \left(\frac{-Q_{SS} - Q_{SD\text{max}}}{\epsilon_{ox}}\right) T_{ox} + \phi_{MS'} + 2\phi_F. \qquad (2.44)$$

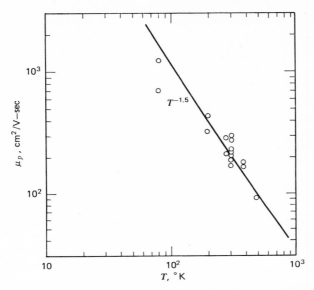

FIGURE 5.2   Effective inversion layer hole mobility as a function of temperature for silicon with $(Q_{total}/q) < 10^{12}/cm^2$. (After Leistiko, Grove, and Sah[1].)

The threshold voltage associated with a MOSFET can easily be obtained experimentally by examining the low-gate-voltage behavior of the device either at low values of drain voltage, well below pinch-off, or within the region of saturated drain current. Below pinch-off, the threshold voltage is obtained in conjunction with (4.20) by measuring the variation of the drain current as a function of the applied gate voltage with the drain voltage held constant and extrapolating the resulting graph to obtain the gate voltage intercept. Alternatively, the threshold voltage of the device may be measured in the saturated drain current region in conjunction with (4.41) by measuring the behavior of the *square root* of the drain current as a function of the applied gate voltage either at constant drain voltage beyond pinch-off or with the gate electrode and the drain electrode tied together. Once again, the extrapolated gate voltage intercept is, by definition, equal to the threshold voltage. When measured through the use of either of the above techniques, the threshold voltage will be independent of the ratio of the channel width to the channel length of the MOSFET.

The fixed positive interface charge density per unit area at the oxide-silicon interface, Q_{SS}, has been shown experimentally to be approximately independent of temperature over a wide range of temperatures.[2,3] If $\phi_{MS'}$ is also assumed to be independent of temperature, differentiation of (2.44) with respect to temperature yields

$$\frac{dV_T}{dT} \cong -\frac{1}{C_{ox}}\left(\frac{dQ_{SD_{max}}}{dT}\right) + 2\left(\frac{d\phi_F}{dT}\right). \tag{5.1}$$

Now, since

$$Q_{SD_{max}} \cong -(4q\epsilon_s N_A \phi_F)^{1/2}, \tag{2.29}$$

the result of the differentiation is

$$\frac{dV_T}{dT} \cong -\frac{1}{C_{ox}}(4q\epsilon_s N_A)^{1/2}\frac{d(\phi_F)^{1/2}}{dT} + 2\frac{d\phi_F}{dT}, \tag{5.2}$$

or

$$\frac{dV_T}{dT} \cong \frac{d\phi_F}{dT}\left[-\frac{Q_{SD_{max}}}{2C_{ox}\phi_F} + 2\right]. \tag{5.3}$$

Again considering the case of an n-channel MOSFET fabricated on a p-type silicon substrate, the Fermi potential associated with the substrate material is given by

$$\phi_F \cong \frac{kT}{q}\ln\left(\frac{N_A}{n_i}\right). \tag{5.4}$$

Differentiating with respect to temperature yields

$$\frac{d\phi_F}{dT} \cong \frac{k}{q}\ln\left(\frac{N_A}{n_i}\right) + \frac{kT}{q}\frac{d}{dT}\left[\ln\left(\frac{N_A}{n_i}\right)\right], \tag{5.5}$$

where n_i, the intrinsic concentration, is given by

$$n_i \cong 3.86 \times 10^{16}T^{3/2}\exp\left(-\frac{E_{GO}}{2kT}\right) \tag{5.6}$$

for silicon at room temperature and above.[4] The constant E_{GO} is approximately equal to 1.21 eV. Substituting (5.6) into (5.5) and differentiating yields

$$\frac{d\phi_F}{dT} \cong \frac{k}{q}\left[\ln\left(\frac{N_A}{n_i}\right) - \frac{E_{GO}}{2kT} - \frac{3}{2}\right]. \tag{5.7}$$

For most temperatures of practical interest,

$$\frac{E_{GO}}{2kT} \gg \frac{3}{2} \tag{5.8}$$

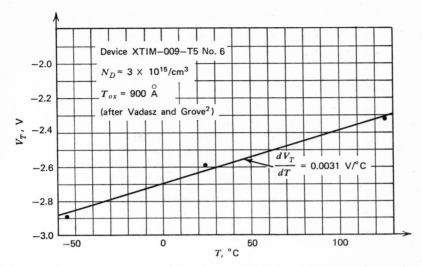

FIGURE 5.3 Variation of threshold voltage with temperature for a p-channel silicon MOSFET. Theoretical behavior obtained through evaluation of (2.44) and (5.10) at room temperature. (After Vadasz and Grove[2].)

and, through the use of (5.4), (5.7) can be expressed as

$$\frac{d\phi_F}{dT} \cong \frac{1}{T}\left[\phi_F - \frac{E_{GO}}{2q}\right]. \tag{5.9}$$

Substitution of (5.9) into (5.3) gives the following expression for the derivative of the threshold voltage with respect to temperature for an n-channel MOSFET fabricated on a p-type substrate:

$$\frac{dV_T}{dT} \cong \frac{1}{T}\left[\left(\phi_F - \frac{E_{GO}}{2q}\right)\left(-\frac{Q_{SD_{max}}}{2C_{ox}\phi_F} + 2\right)\right]. \tag{5.10}$$

A similar expression for p-channel devices fabricated on n-type substrates can easily be derived by analogy. The derivative of the threshold voltage with respect to temperature will always be negative for n-channel MOSFETs and will always be positive for p-channel MOSFETs.

Experimentally, it has been demonstrated that V_T is approximately a linear function of temperature for both n- and p-channel devices over the range -55 to $+125°C$ and its value can be fairly accurately predicted by evaluating (2.44) and (5.10) at room temperature.[2] The correlation between experiment and theory through the use of this technique for a typical p-channel MOSFET is shown in Figure 5.3. The combined effect of the

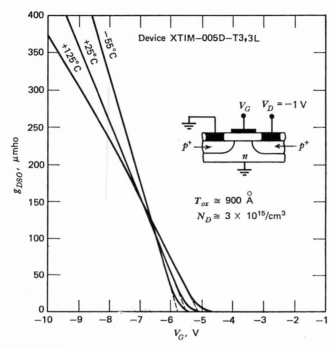

FIGURE 5.4 Variation of the drain-to-source conductance at zero drain voltage with applied gate voltage as a function of temperature for a p-channel MOSFET. (After Vadasz and Grove[2].)

variations with temperature of both β and V_T on the characteristics of a similar p-channel device operating at extremely low drain voltages is shown in Figure 5.4. The drain-to-source conductance with zero drain voltage, g_{DS0}, is plotted versus the applied gate voltage as a function of temperature. The g_{DS0} is defined by

$$g_{DSO} \equiv \left(\frac{\partial I_D}{\partial V_D}\right)_{V_D=0} = |\beta(V_G - V_T)|. \tag{5.11}$$

The reduction in the overall gain of the device with increasing temperature can be seen from the steadily decreasing slope of the conductance-versus-gate voltage curves, while the decrease in the observed negative threshold voltage with increasing temperature can be seen from the change in the gate voltage intercept.

Qualitatively, a similar behavior with temperature variation will be observed when a MOSFET is operated beyond pinch-off in the region of saturated

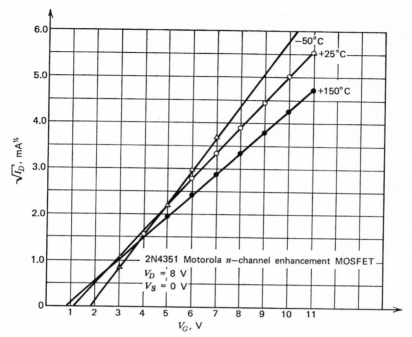

FIGURE 5.5 Variation of the square root of the saturated drain current beyond pinch-off versus the applied gate voltage as a function of temperature for an *n*-channel enhancement MOSFET.

current flow,[5] as in Figure 5.5, where the square root of the saturated drain current is plotted versus the applied gate voltage with the drain voltage held constant for an *n*-channel enhancement-type device.

It should be noted that, unlike *p*-channel MOS digital integrated circuits, *n*-channel circuits usually are operated with applied substrate potential, since *n*-channel MOSFETs that are fabricated on relatively-high-resistivity *p*-type silicon substrates are usually of the *depletion type* and must be biased in this manner to achieve enhancement-type characteristics. Since the maximum charge density in the surface depletion region per unit area will be a function of the applied substrate-to-source voltage, it follows that the rate of change of the threshold voltage with temperature will also be modified by the application of the substrate potential. Under these conditions, from Section 2.3, the threshold voltage of an *n*-channel MOS transistor will be approximately given by

$$V_T \cong \left(\frac{-Q_{SS} - Q'_{SD_{max}}}{\epsilon_{ox}} \right) T_{ox} + \phi_{MS'} + 2\phi_F, \qquad (5.12)$$

where $Q'_{SD_{max}}$ is given by (2.78). Again, assuming that the quantities Q_{SS} and $\phi_{MS'}$ are independent of temperature, the variation with temperature of the threshold voltage can be obtained by differentiating (5.12):

$$\frac{dV_T}{dT} \cong -\frac{dQ'_{SD_{max}}}{dT}\left(\frac{T_{ox}}{\epsilon_{ox}}\right) + 2\frac{d\phi_F}{dT}. \tag{5.13}$$

Differentiation with respect to temperature of the maximum value of the charge density per unit area in the surface depletion region with zero substrate-to-source voltage, as given by (2.29), yields

$$\frac{dQ_{SD_{max}}}{dT} \cong \left(\frac{Q_{SD_{max}}}{2\phi_F}\right)\frac{d\phi_F}{dT}, \tag{5.14}$$

and, from (2.78), it follows that

$$\frac{dQ'_{SD_{max}}}{dT} \cong \frac{d}{dT}\left[Q_{SD_{max}}\left(\frac{2\phi_F - V_S}{2\phi_F}\right)^{1/2}\right], \tag{5.15}$$

or

$$\frac{dQ'_{SD_{max}}}{dT} \cong Q_{SD_{max}}\frac{d}{dT}\left(\frac{2\phi_F - V_S}{2\phi_F}\right)^{1/2} + \left(\frac{2\phi_F - V_S}{2\phi_F}\right)^{1/2}\left(\frac{dQ_{SD_{max}}}{dT}\right). \tag{5.16}$$

Using (2.78), (5.13), (2.29), (5.14), and (5.16) and rearranging yields an expression for the rate of change of threshold voltage with respect to temperature:

$$\frac{dV_T}{dT} \cong \frac{d\phi_F}{dT}\left\{2 - \frac{T_{ox}}{\epsilon_{ox}}\left\{\frac{Q_{SD_{max}}V_S}{4\phi_F^2[1 - (V_S/2\phi_F)]^{1/2}} + \left(1 - \frac{V_S}{2\phi_F}\right)^{1/2}\left(\frac{Q_{SD_{max}}}{2\phi_F}\right)\right\}\right\}. \tag{5.17}$$

Equation 5.17 can be greatly simplified, and reduces to the following:

$$\frac{dV_T}{dT} \cong \frac{d\phi_F}{dT}\left\{2 - \frac{T_{ox}Q_{SD_{max}}}{2\epsilon_{ox}\phi_F[1 - (V_S/2\phi_F)]^{1/2}}\right\}, \tag{5.18}$$

or

$$\frac{dV_T}{dT} \cong \frac{d\phi_F}{dT}\left\{2 - \frac{Q_{SD_{max}}}{2C_{ox}\phi_F[1 - (V_S/2\phi_F)]^{1/2}}\right\}. \tag{5.19}$$

Equation 5.19 is valid for the case of an n-channel MOSFET fabricated on a p-type substrate (with ϕ_F positive and V_S less than or equal to 0 V). Once again, a similar expression can be derived by analogy for the case of a p-channel MOSFET fabricated on an n-type substrate (with ϕ_F negative and

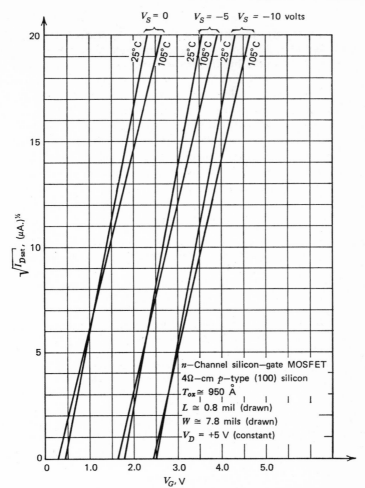

FIGURE 5.6 Variation of the square root of the saturated drain current with gate voltage for two different temperatures as a function of the applied substrate-to-source voltage for an *n*-channel silicon gate MOSFET.

V_S greater than or equal to 0 V). As can be seen from the equations above, the effect of *increasing* the applied substrate-to-source potential is to *decrease* the rate of change of the threshold voltage with temperature. When the applied substrate-to-source voltage is set equal to zero, it should be noted that (5.19) reduces to (5.3).

Experimental verification of the fact that, as predicted by (5.19), the rate of change of the threshold voltage with temperature decreases with increasing applied substrate-to-source voltage can be seen from the data shown in Figure 5.6. Qualitatively, the difference between the extrapolated values of threshold voltage at $+105$ and $+25°C$ of an n-channel enhancement-type silicon-gate MOSFET operating under saturated drain current conditions is observed to decrease with increasing reverse substrate bias. Quantitatively, the resulting experimental values of the rate of change of the threshold voltage with respect to temperature as a function of the substrate bias for the device shown are actually slightly lower than the theoretical values predicted by (5.19). Wang, Dunkley, DeMassa, and Jelsma[6] have shown that a better correlation between theory and experiment in this case can be obtained by considering the behavior of $\phi_{MS'}$ with temperature and not assuming $(d\phi_{MS'}/dT)$ to be equal to zero.

REFERENCES

1. O. Leistiko, Jr., A. S. Grove, and C. T. Sah, Electron and Hole Mobilities in Inversion Layers on Thermally Oxidized Silicon Surfaces, *IEEE Transactions on Electron Devices*, Vol. ED-12, No. 5, 1965, pp. 248–254.

2. L. Vadasz and A. S. Grove, Temperature Dependence of MOS Transistor Characteristics Below Saturation, *IEEE Transactions on Electron Devices*, Vol. ED-13, No. 12, 1966, pp. 863–867.

3. A. S. Grove, B. E. Deal, E. H. Snow, and C. T. Sah, Investigation of Thermally Oxidized Silicon Surfaces Using Metal-Oxide-Semiconductor Structures, *Solid State Electronics*, Vol. 8, 1965, pp. 145–163.

4. F. J. Morin and J. P. Maita, Electrical Properties of Silicon containing Arsenic and Boron, *Physical Review*, Vol. 96, No. 1, October 1954, pp. 28–35.

5. R. S. C. Cobbold, Temperature Effects on MOS Transistors, *Electronics Letters*, Vol. 2, No. 6, June 1966, pp. 190–191.

6. R. Wang, J. Dunkley, T. DeMassa, and L. Jelsma, Threshold Voltage Variations with Temperature in MOS Transistors, *IEEE Transactions on Electron Devices*, Vol. ED-18, No. 6, 1971, pp. 386–388.

BIBLIOGRAPHY

Degraaff, H. C., and J. A. V. Nielen, Temperature Influence on the Channel Conductance of MOS Transistors, *Electronics Letters*, Vol. 3, May 1967, pp. 195–196.

Gray, P. V., and D. M. Brown, Freeze-Out Characteristics of the MOS Varactor, *Applied Physics Letters*, Vol. 13, No. 8, October 1968, pp. 247–248.

Heiman, F. P., and H. S. Miller, Temperature Dependence of n-type MOS Transistors, *IEEE Transactions on Electron Devices*, Vol. ED-12, No. 3, March 1965, pp. 142–148.

Nathanson, H. C., C. Jund, and J. Grosvalet, Temperature Dependence of Apparent Threshold Voltage of Silicon MOS Transistors at Cryogenic Temperatures, *IEEE Transactions on Electron Devices*, Vol. ED-15, No. 6, June 1968, pp. 362–368.

Rogers, C. G., MOST's at Cryogenic Temperatures, *Solid State Electronics*, Vol. 11, 1968, pp. 1079–1091.

Sesnic, S. S., and G. R. Craig, Thermal Effects in JFET and MOSFET Devices at Cryogenic Temperatures, *IEEE Transactions on Electron Devices*, Vol. ED-19, No. 8, August 1972, pp. 933–942.

Valkof, S. A., Temperature Stabilization of MOS Transistor Gain, *Proceedings of the IEEE*, Vol. 59, No. 9, 1971, pp. 1374–1375.

Zuleeg, R., Temperature Compensation Effect in MOS Transistors, *Proceedings of the IEEE*, Vol. 53, No. 7, 1965, pp. 732–734.

PROBLEMS

5.1 Derive an expression for the derivative of the threshold voltage of a p-channel MOSFET with respect to temperature in terms of T, ϕ_F, $Q_{SD_{\max}}$, V_S, and C_{ox}.

5.2 Discuss the effect of temperature variation on the electrical characteristics of MOSFETs as a function of substrate doping concentration. In particular, discuss the behavior of β and V_T as a function of temperature for both low-resistivity and high-resistivity substrates.

5.3 Discuss the possible significance of the observed $T^{-3/2}$ dependence of the high-temperature field-effect carrier mobility. Which scattering mechanism seems to be dominant?

5.4 Plot the theoretical value of the strong-inversion threshold voltage V_T as a function of temperature from -55 to $+125°C$ for an n-channel MOSFET fabricated on a p-type silicon substrate with $N_A = 10^{15}/cm^3$ through the use of (2.44). Assume that aluminum is used for the gate electrode, the thickness of the silicon dioxide gate insulator is equal to 1000 Å, and that $Q_{SS}/q = 1.5 \times 10^{11}/cm^2$.

5.5 Evaluate both V_T and (dV_T/dT) at 25°C for the device described in problem 5.4. Approximate V_T as a linear function of temperature from -55 to $+125°C$ by extrapolating the slope obtained from the room temperature value of (dV_T/dT) through the room temperature value of the threshold voltage. Compare this linear approximation with the actual behavior of V_T as a function of temperature as obtained in problem 5.4.

5.6 Using (2.78), (5.13), (2.29), (5.14), and (5.16), derive the expression given in (5.17).

5.7 Demonstrate that (5.17) can be simplified into the form given in (5.19).

5.8 Discuss physically why the application of a reverse substrate-to-source bias should decrease the rate of change of the threshold voltage with respect to temperature.

5.9 Repeat problem 5.4 for the case when the substrate-to-source voltage is not equal to zero and, in particular, when $V_S = -2.5$, -5.0, and -10.0 V.

6

The Silicon-Silicon Dioxide System

Although insulated-gate field-effect transistors have been fabricated using a wide variety of combinations of semiconductor substrates and gate insulators, until now only silicon MOS devices have achieved widespread acceptance and use throughout the microelectronics industry, because of two very important considerations:

1. The ability of silicon to be thermally oxidized to form an insulator.
2. The compatibility of the silicon MOS device with silicon planar processing techniques for integrated circuit fabrication.

Because of the tremendous importance of the silicon MOSFET, virtually all of the work that has been done so far in the study of insulator-semiconductor interface properties has been devoted to the silicon-silicon dioxide system. Since the operation of the MOSFET is based on the formation of a conducting channel at the surface of the silicon substrate, it would seem physically obvious that the characteristics and operation of the device would depend heavily on the nature of the surface. Furthermore, since the transition at the interface from semiconductor to insulator is so drastic, it might be expected that many of the bulk parameters normally used to characterize the silicon would require extensive modification to describe more accurately the actual characteristics of the device.

This chapter examines briefly the properties of the silicon-silicon dioxide system and, in particular, the characteristics of the silicon-silicon dioxide interface, and the resulting effects on the properties of silicon MOS devices.

137

6.1 SURFACE STATES AND THEIR INFLUENCE ON THE ELECTRICAL CHARACTERISTICS OF MOS DEVICES

Early studies of the silicon-silicon dioxide system showed that the characteristics of the oxidized semiconductor were strongly influenced, and at times dominated, by the properties of the silicon-silicon dioxide interface. It became readily apparent that the discrepancies found between experimental results obtained with MOS capacitors and transistors and the early theoretical treatments based on an ideal metal-insulator-semiconductor structure could only be reconciled by postulating the existence of an interface charge density, the characteristics of which were strongly dependent on the procedures used in fabricating the device and, quite often, were also dependent on the applied voltages. In addition, further research demonstrated the existence of both fixed and mobile ionic charge within the silicon dioxide insulator, a great deal of which was shown to be in close proximity to the silicon-silicon dioxide interface, thereby having a strong effect on the observed device characteristics.

As a result of the enormous amount of research done over many years, a thorough understanding of the silicon-silicon dioxide system has been obtained which can be used to explain virtually all the observed electrical characteristics of silicon MOS devices. These characteristics have been shown to be consistent with the existence of the following:

1. "Fast" surface states located at the silicon-silicon dioxide interface.
2. A fixed *positive* surface-state charge density located at or near the interface.
3. Both fixed and mobile ionic charge located within the silicon dioxide, resulting either from ionizing radiation or (sodium) contamination during device fabrication.
4. Impurity redistribution at the silicon surface during thermal oxidation.
5. A number of different instability mechanisms.

The following sections are devoted to discussing the phenomena above and, in particular, to describing each of their effects on the characteristics of MOSFETs.

6.1.1 Fast Surface States

Surface states located at the silicon-silicon dioxide interface with energies falling within the silicon bandgap are commonly referred to as "fast surface states," because the charge located in these states can readily and rapidly be exchanged with the silicon substrate. When a voltage is applied across a MOS capacitor, any variation in the applied gate-to-substrate potential will result in a change in the amount of surface band-bending relative to the

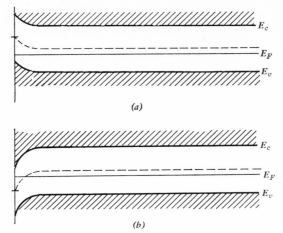

FIGURE 6.1 Occupancy of fast surface states at the silicon-silicon dioxide interface as a function of the amount of surface band-bending.

Fermi level in the silicon, as is shown in Figure 6.1. Thus if there are fast surface states located at the interface, the probability of their being occupied will vary as a function of the applied bias. As shown in Figure 6.1a, if one considers the case of fast surface states with energies located at or near the middle of the bandgap, when a negative voltage is applied to the gate of a MOS capacitor formed on a p-type silicon substrate, the energy bands bend upward at the surface as the silicon becomes accumulated and the position of the Fermi level relative to the energy of the states is such that their probability of occupancy is relatively small. However, as shown in Figure 6.1b, if a positive voltage is applied to the gate electrode of the structure, the silicon surface becomes inverted and the position of the Fermi level is now such that the probability of electrons being trapped by the fast surface states is extremely high. Therefore, it can be easily seen that the amount of charge located in the fast surface states at the interface will be a function of the surface potential and, consequently, will vary with the applied gate voltage. The total amount of charge located in surface states will also depend greatly on their distribution in energy throughout the bandgap and whether they are acceptor or donor types. Nicollian and Goetzberger[1] and Gray and Brown[2,3] have reported data that indicate that the distribution of fast surface states across the silicon bandgap typically exhibits peaks near both band edges with a pronounced and uniform minimum occurring near the center of the forbidden region. This is illustrated in Figure 6.2 for a MOS capacitor structure with a silicon dioxide gate insulator that was formed by thermal oxidation of silicon in a dry oxygen environment at 1200°C.[4]

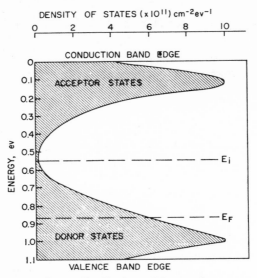

FIGURE 6.2   Approximate distribution of fast surface states associated with the silicon-silicon dioxide interface following a rapid quench into room ambient from a 1200°C dry oxygen oxidation. (After Castro and Deal[4].)

In general, the origin of fast surface states, as is the case with the fixed positive interface charge density Q_{SS}, has usually been linked with a disruption of the periodicity of the silicon lattice near the silicon-silicon dioxide interface. In particular, it has been proposed that the thermal oxidation of silicon may give rise to an excess of silicon ions located in the silicon dioxide in close proximity to the interface, resulting in an observed positive charge density (Q_{SS}) whose magnitude is independent of the surface potential. Furthermore, it has been suggested that the movement of these silicon ions into the oxide may result in vacancies in the silicon occurring at or near the surface that, in turn, might lead to an increase in the fast surface state density and, consequently, lead to an increase in surface recombination rates.[5]

It is well-known that the density of fast surface states at the silicon-silicon dioxide interface can be reduced to extremely low levels by annealing in an inert environment at low temperatures if an aluminum gate electrode is present directly over the silicon dioxide.[4] The reduction in fast surface state density resulting from the low-temperature inert-atmosphere annealing of aluminum-gate MOS structures has been attributed to the annihilation of these states by an active hydrogen species which is formed by the reaction of residual traces of moisture in the oxide with the aluminum electrode during the anneal. In structures where the aluminum gate is not present (e.g., silicon-gate devices) the fast surface state density can be effectively reduced by annealing either in pure hydrogen or in forming gas at temperatures ranging from 400 to 500°C.[4,6] Oxides grown in dry oxygen are characterized by relatively high fast surface state densities, while even trace amounts of water present during oxidation will reduce the fast state density to levels well below 10^{10}/cm². Even if it is kept low because of oxidation in the presence of water vapor, the fast surface state density will still increase substantially if the water is subjected to a subsequent anneal in an inert environment.[7] However, the formation of fast surface states during oxidation or annealing in an inert environment is not considered important because of the previously discussed ease with which these states can be subsequently annihilated during the final processing steps associated with both aluminum-gate and silicon-gate structures.

Gray and Brown[2] have found that, with all other material and processing conditions held constant, the observed density of fast surface states is greatest on (111) orientation silicon and is smallest on (100) silicon. The values observed on (110) material typically lie somewhere in between. It is interesting to note that this same type of behavior is found for the fixed positive interface charge density per unit area, Q_{SS}. That is, the observed magnitude of Q_{SS} as a function of silicon substrate orientation decreases in the order (111) > (110) > (100).[5] Although the fixed positive interface charge density and the fast surface state density in many ways behave quite similarly, it should be remembered that they are distinctly different quantities and their respective densities can be varied independently through appropriate process variations.[7]

In general, the presence of fast surface states has a degrading effect on the electrical characteristics of MOSFETs and, when possible, the fast state density should be reduced to a minimum level. Since they can easily exchange charge with the silicon substrate, fast surface states can act as surface recombination centers. It has been found that the fast states located near the center of the bandgap are primarily responsible for surface recombination and can cause a considerable increase in the leakage current observed at the drain junction.[7] Also, fast states act as scattering centers at the silicon-silicon dioxide interface, thereby reducing the effective carrier mobility in the channel. They can also reduce the observed field-effect mobility by immobilizing a portion of the carriers that would otherwise contribute to channel conductivity. Similarly, this effect can also contribute to an increase in the observed threshold voltage. Fast surface states have also been shown to be responsible for the low-frequency ($1/f$) noise typically found in MOS devices. The effect of fast surface states on the capacitance-versus-voltage characteristics of MOS capacitors is treated in Section 3.3.

6.1.2   The Fixed Positive Interface Charge Density, Q_{SS}

In general, moderately doped p- and n-type silicon have been observed to have a distinct tendency to form, respectively, n-type surface inversion or accumulation layers when thermally oxidized. The reason for this behavior has been traced by many researchers to the existence of a fixed positive charge density that is located near the silicon-silicon dioxide interface. This surface state charge density, which is denoted, per unit area, by Q_{SS}, differs greatly from the fast surface state charge density in that its magnitude, for all practical purposes, is *not* a function of the applied gate voltage or the surface potential in the silicon near the interface because the energy levels of the states associated with the fixed positive interface charge density, unlike the energy levels associated with fast surface states, lie outside of the forbidden bandgap, so that their probability of occupancy will not be affected as the Fermi level sweeps across the forbidden band with variations in the applied gate voltage. Revesz, Zaininger, and Evans[8] have found that the Q_{SS} charge density results from the presence of a distribution of donor-like states located higher than about 0.8 eV above the silicon valence band edge (and probably even above the conduction band edge) that are permanently ionized.

The basic characteristics of the fixed positive interface charge density Q_{SS} were described and discussed in a classic paper by Deal, Sklar, Grove, and Snow.[5] They found it to have the following properties:

1. Its magnitude is fixed and cannot be varied by band-bending of the silicon energy bands over at least the middle 0.7 eV of the forbidden energy gap.

2. Its magnitude is not significantly influenced by the thickness of the overlying silicon dioxide, nor is it significantly affected by the type or concentration of doping impurities in the silicon substrate in the range 10^{14} to $10^{17}/cm^3$.

3. Its density is stable under elevated temperature-applied bias test conditions up to about 250°C, which would normally cause movement of mobile positive ionic charges located in the silicon dioxide gate insulator. Thus it is not in any way related to sodium and other species of mobile ionic contamination that might be introduced during device fabrication.

4. It is located in the silicon dioxide, in close proximity (0 to 200 Å) to the silicon-silicon dioxide interface.

5. Its magnitude, under similar processing conditions, is a strong function of the crystalline orientation of the silicon substrate. As discussed in the previous section, the observed magnitude of Q_{SS} as a function of silicon substrate orientation under similar processing conditions decreases in the order (111) > (110) > (100) in approximately the ratio 3:2:1.

6. Its magnitude can be greatly influenced by changes in the ambient

(i.e., water vapor or dry oxygen) and temperature during thermal oxidation and can be reproducibly varied during subsequent high-temperature annealing treatments in different ambients.

The last characteristic of the fixed positive interface charge density per unit area mentioned above is particularly interesting and is of great importance in specifying and controlling the final electrical characteristics of any silicon MOSFET. Deal, Sklar, Grove, and Snow found that, while the magnitude of Q_{SS} for oxides grown in the presence of water vapor varied only slightly as a function of the oxidation temperature over a wide range (decreasing slightly with increasing temperature), this was not true of oxides grown in a dry ambient. In particular, they found that the value of Q_{SS} associated with oxides grown in a dry O_2 ambient decreased fairly rapidly as the oxidation temperature was increased over the range from 700 to 1200°C. Furthermore, they showed that once the oxide layer had been grown, the value of Q_{SS} could be reproducibly changed by either annealing for a short time in dry oxygen at a different temperature or by annealing in a dry inert gas. When the latter technique was employed, it was found that the magnitude of Q_{SS} was always reduced to a minimum level which was approximately independent of the temperature of the dry inert anneal. The time required for the Q_{SS} value to approach a constant equilibrium level during the various oxidation and annealing treatments was found to vary from about an hour at 550°C to less than 10 mins at 900°C for oxide thicknesses on the order of 2000 Å. The experimental results of the work above are summarized in Figure 6.3. The data were obtained from MOS capacitors fabricated on both p- and n-type (111) silicon substrates with impurity doping concentrations of approximately $1.4 \times 10^{16}/cm^3$ and silicon dioxide gate insulators approximately 2000 Å thick.[5] Data obtained on (100) orientation silicon substrates show a similar dependence, but the values of Q_{SS} observed are substantially lower.

While the behavior above may vary quantitatively from system to system depending on the type of oxidation equipment employed, the qualitative behavior is found to be very reproducible. It is indeed important to note that *the only relevant high-temperature treatment in setting the Q_{SS} level is the final one.* Thus for example, low Q_{SS} levels are typically obtained by growing the gate insulator in dry oxygen at any temperature and then subjecting the wafer to an inert high-temperature anneal in dry nitrogen. If a particular level of Q_{SS} is desired (above the minimum level) it can be achieved by either growing the gate insulator in dry oxygen at the desired temperature or by growing the oxide in either a wet or dry oxygen ambient at any temperature and subsequently annealing in dry oxygen at the desired temperature required to specify the Q_{SS} level.

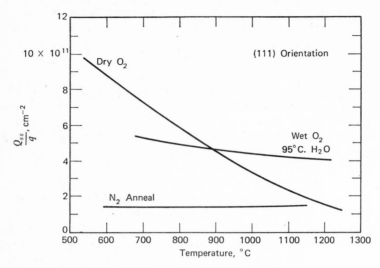

FIGURE 6.3 Dependence of the fixed positive interface charge density per unit area Q_{SS} on the ambient and temperature of the final heat treatment. (After Deal, Sklar, Grove, and Snow[5].)

As discussed in Section 3.3, the effect of Q_{SS} is to displace the capacitance-versus-voltage characteristic associated with a MOS device in the negative direction without changing its shape. Although the fixed positive charge density per unit area Q_{SS} will thus greatly affect the observed flatband and threshold voltages of MOS capacitors and transistors, respectively, it will not have any serious effects on most other important electrical characteristics.

6.1.3   The High-Temperature Negative Bias Instability

Although the values of the fixed positive interface charge density per unit area typically observed for MOS devices have been shown to be relatively independent of temperature over the range of -55 to $+125°C$ and, beyond that, even up to $+250°C$, it has been discovered that the apparent Q_{SS} level can be shifted when bias is applied to the gate electrode of a MOS device which is operating in an ambient in excess of $300°C$.[5,7,9] Specifically, an increase in the apparent Q_{SS} level is observed when a strong *negative* bias is applied to the gate electrode, thereby resulting in a negative shift in the flatband voltage. The subsequent application of a positive voltage of the same magnitude to the gate electrode will result in partial recovery, but the amount of positive shift in the flatband voltage at a given temperature above $300°C$ will not be nearly as great as the original negative shift at that temperature.

This phenomenon is commonly referred to as the *high-temperature negative bias instability*. Deal, Sklar, Grove, and Snow[5] have shown that the amount of negative shift of the flatband voltage associated with this instability is linearly proportional to the electric field strength applied across the gate insulator and is also proportional to the (original) value of Q_{SS} observed previously at low temperatures. Thus the high-temperature negative-bias instability has a much greater effect, for example, on silicon substrates with (111) orientation than on substrates with (100) orientation. The total amount of flatband shift associated with the instability was found to saturate rapidly at 400°C, taking approximately 1 to 3 min, while saturation took considerably longer at lower temperatures (typically about 8 hr at 300°C). It is also important to note that Deal, Sklar, Grove, and Snow found that the increase in Q_{SS} associated with the application of a negative gate voltage at temperatures greater than 300°C also generally resulted in a simultaneous increase in the fast state density of approximately the same magnitude, and that the combined effects of the changes in both quantities could result in very large shifts and gross distortions in the shape of the observed capacitance-versus-voltage curves.

6.2 IMPURITY REDISTRIBUTION IN SILICON DURING THERMAL OXIDATION

Early in 1960, Atalla and Tannenbaum found that the impurity doping concentration at the surface of a silicon substrate was altered by the effects of thermal oxidation at elevated temperatures.[10] Subsequent work by other researchers showed a wide variation in the extent and type of redistribution during oxidation exhibited by different impurities commonly used to dope silicon.[11,12] Obviously, those electrical parameters of a MOS device which are in any way dependent on the surface doping concentration in the silicon substrate will be affected by the impurity redistribution process.

Impurity redistribution is certainly not unique to the silicon-silicon dioxide system. If any two materials are brought together, a small impurity contained (e.g., in solution) in one will tend to be distributed gradually into the other until equilibrium is reached when the ratio of the concentrations in the two phases becomes constant. The amount of redistribution and the rate at which the redistribution takes place will be a function of many variables, one of the most important being the ratio above of the equilibrium impurity concentrations between the two phases. This ratio is called the *segregation coefficient* and, for the silicon-silicon dioxide system, is defined by

$$m \equiv \frac{\text{equilibrium concentration of the impurity in silicon}}{\text{equilibrium concentration of the impurity in the silicon dioxide}}.$$

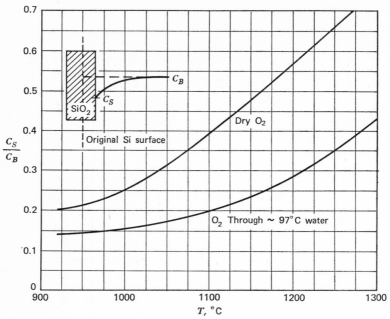

FIGURE 6.4   Redistribution of boron during thermal oxidation; $C_B \sim 10^{16}/cm^3$. (After Grove[13].)

Boron, for example, has been shown to exhibit a segregation coefficient of approximately 0.3 while phosphorus exhibits a much larger value, on the order of 10.[11,12] Another important factor in the redistribution process is the fact that thermally grown silicon dioxide takes up a much greater volume than the amount of silicon which is consumed during the oxidation process. (Approximately 50 to 55% of the silicon dioxide that is grown lies above the original silicon surface.) Thus the impurities which move into the silicon dioxide as the silicon surface is being oxidized must be distributed into a larger volume, which tends to result in depletion of the impurity at the new silicon-silicon dioxide interface. The rate of diffusion of the impurity in the silicon dioxide will also affect greatly the impurity redistribution process.

A theory for impurity redistribution at the surface of a silicon substrate during thermal oxidation which is consistent with experimentally observed results has been developed by Grove, Leistiko, and Sah.[11] The theory predicts that boron will tend to be depleted from the silicon surface during thermal oxidation while phosphorus will tend to pile-up at the silicon surface. The theory also indicates that, under equilibrium conditions, the impurity concentrations at the silicon-silicon dioxide interface will be *independent of*

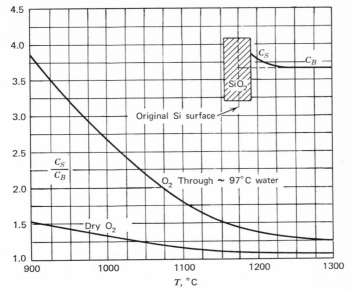

FIGURE 6.5 Redistribution of phosphorus during thermal oxidation; $C_B \sim$ $10^{16}/cm^3$. (After Grove[13].)

oxidation time, and will only vary with oxidation temperature. The depth of the redistribution into the silicon and the resulting impurity profile will be a strong function of both the oxidation temperature and the time required to complete the oxidation. (The longer the oxidation time at a given temperature, the deeper the effects of the redistribution process.)

The effects of impurity redistribution on the surface concentrations of boron-doped and phosphorus-doped silicon substrates are shown in Figures 6.4 and 6.5, respectively, as a function of oxidation temperature.[13] In each case, the impurity doping concentration deep in the bulk of the substrate is on the order of $10^{16}/cm^3$. Note that the ratio of the surface concentration to the concentration deep in the substrate is always greater, for all temperatures, *for boron-doped substrates*, when the oxidation is performed in a dry ambient than when the oxidation takes place in wet oxygen. On the other hand, the ratio of the surface concentration to the concentration deep in the substrate is always greater, for all temperatures, for wet oxidation than for dry oxidation if the silicon substrate is *phosphorus-doped*. The behavior above is consistent with the fact that the amount of surface impurity redistribution will increase as the oxidation rate at a given temperature is increased. Consequently, the effect of increasing the oxidation rate by oxidizing in a wet ambient rather than in dry oxygen will be to create either a greater depletion of boron at the surface of a boron-doped wafer or a greater accumulation of phosphorus at the surface of a phosphorus-doped wafer. The amount of impurity redistribution of both boron and phosphorus *at the silicon surface* can be minimized by oxidizing at higher temperatures. However, in the process of growing a predetermined thickness of silicon dioxide, the *depth* of the redistribution into the silicon will be larger for higher oxidation temperatures.

In general, the effects of impurity redistribution during thermal oxidation on the electrical parameters of typical silicon MOS devices are relatively small. Schottky[14] has shown, however, that the depletion of boron from the silicon surface during thermal oxidation can have a substantial effect on the threshold voltage of an *n*-channel MOSFET which is operating with applied substrate-to-source voltage. For typical substrate doping concentrations used in the manufacture of MOS devices, the effect of either boron depletion or phosphorus "pile-up" or accumulation upon the threshold voltage with zero substrate-to-source bias is small. However, as is typically the case with *n*-channel MOSFETs that are operated with applied substrate voltage in order to set their threshold voltages at a predetermined value, when a reverse substrate bias is applied, the effect of the depletion charge density term in the equation for the threshold voltage is greatly increased and the effect of impurity redistribution on this term and, correspondingly, on the threshold voltage, can no longer be considered negligible. (See Section 2.3.) While *p*-channel MOSFETs usually do not require any substrate-to-source bias, a negative substrate voltage is usually required to achieve enhancement-type operation of *n*-channel devices for digital logic applications. Under these conditions, Schottky has shown that the effect of boron depletion in the channel region can shift the observed threshold voltage in the negative direction by as much as 1 V.[14]

6.3 THE EFFECTS OF RADIATION-INDUCED OXIDE CHARGE AND SURFACE STATES ON THE CHARACTERISTICS OF MOS TRANSISTORS

The MOSFET and related devices have been shown to be quite susceptible to appreciable changes in their electrical characteristics resulting from γ-ray, x-ray, and electron irradiations. The overall effects of such ionizing radiation upon MOS devices can be classified into two major categories:[15–18]

1. The build-up of a positive space-charge region within the silicon dioxide gate insulator when the gate electrode of the device is biased at either a positive *or* a negative potential.

2. An increase in the density of fast surface states at the silicon-silicon dioxide interface.

The positive space-charge that can build-up in the gate insulator will result in a negative shift of the flatband voltage of a MOS capacitor, and in a negative shift of the threshold voltage of a MOS transistor along with a corresponding shift in its drain current versus gate voltage characteristics. The creation of fast surface states can, as discussed in Section 6.1.1, result not only in a change in the threshold voltage of a MOSFET, but also in excess diode leakage current and in a reduction of the carrier mobility in the channel region. Although all of these effects can easily be annealed out either in a dry nitrogen ambient (for aluminum-gate devices with silicon dioxide gate insulators) or in a hydrogen ambient at temperatures usually in excess of 400°C, it is obvious that ionizing radiation can have a seriously degrading effect on a MOS device that is operating in an electronic system. Consequently, a great deal of research is being performed in an effort to reduce the degrading effects of ionizing radiation upon MOS device characteristics.

The formation of the positive space-charge region in the gate insulator of a MOS device operating with a nonzero applied gate-to-source voltage has been shown to be explainable in terms of the following model:

When a MOS device fabricated with a silicon dioxide gate insulator is irradiated, the incident radiation ionizes localized regions in the gate insulator by creating hole-electron pairs within the silicon dioxide. It is generally accepted that the mobility of holes in silicon dioxide is much less than the electron mobility under similar measuring conditions. Thus for an applied gate voltage of *either* polarity, the electrons tend to drift toward the positive electrode and are eventually swept out of the gate insulator, while the less-mobile holes remain behind and soon become trapped in localized sites within the silicon dioxide. Under these conditions, if there is no injection of electrons from the negative electrode, a positive space-charge region will begin to form in the gate insulator near the negative electrode. A steady-state equilibrium condition will result when the applied gate-to-substrate voltage is totally dropped across the positive space-charge region and, as a result, the electric field in the remainder of the insulator will go to zero. The electrons remaining in the latter region that have not been swept out of the silicon dioxide will tend to recombine with the holes there. Although it might be expected that, for zero gate-to-substrate bias, there would be no shift in either the flatband or threshold voltage after irradiation, a small amount of shift is typically observed. This shift, however, can be traced to the work function difference between the gate electrode and the silicon substrate that results in a small "built-in" electric field across the gate insulator, even with zero bias.

The observed threshold voltage shift increases with the magnitude of the applied gate bias and is larger with positive gate voltage than with negative gate voltage because the formation of a positive space-charge layer in close proximity to the silicon-silicon dioxide interface will have a much greater effect on the surface band-bending than if the space-charge region were to form near the gate electrode. The resulting shift in threshold voltage or flatband voltage measured after irradiation with applied gate bias is permanent at room temperature.[18]

In general, the attempts made to reduce the effects of ionizing radiation on the electrical characteristics of MOSFETs fall into two categories: (1) the use of other insulators or combinations of insulating layers to form the gate dielectric, and (2) attempts to match more closely the effective mobilities of the holes and electrons in silicon dioxide. Zaininger and Waxman have reported that MOS devices fabricated with Al_2O_3 gate insulators (which have been grown by means of plasma anodization of a thin layer of aluminum) exhibit electrical characteristics that are remarkably unaffected by incident radiation under normal operating conditions.[17] In particular, they found only very slight shifts in the flatband and threshold voltages of the devices because of space-charge formation in the aluminum oxide compared to similar devices fabricated with silicon dioxide gate insulators. Donovan and Simons, on the other hand, found that the magnitude of the observed positive space-charge build-up in the silicon dioxide gate insulator of a MOS device that was subjected to incident ionizing radiation could be greatly reduced through the use of ion-implantation techniques.[19] They demonstrated that MOS capacitors fabricated with silicon dioxide insulators which were exposed to a nitrogen ion beam with energies in the 50 to 200 keV range seemed to exhibit less space-charge formation after irradiation than similar nonimplanted devices. This behavior was attributed to the mobility-lifetime product of the electrons more closely matching the mobility-lifetime product of holes in silicon dioxide because of a change in the defect structure of the insulator resulting from the implantation. The implantation seemed to have little effect on the creation of interface states by the ionizing radiation.

6.4   IONIC CONDUCTION WITHIN THE GATE INSULATOR

As was shown in Section 2.2.3, the effect of a given amount of charge within the gate insulator on the electrical characteristics of a MOS device is strongly dependent on the spatial distribution of the charge with respect to the silicon-silicon dioxide interface. Specifically, charge located in close proximity to the interface will tend to induce an equal an opposite amount of charge in the silicon near the surface, thereby greatly influencing the threshold or flatband voltage of the device, while charge located near the gate electrode will have little effect on the silicon surface and, consequently, on the observed electrical characteristics. It follows that if any of the ionic charge located within the gate insulator is free to move when acted upon by a strong electric field, substantial changes in the electrical characteristics of the device will be observed as a function of time with the application of a constant d.c. gate bias.

Hydrogen ions and, in particular, sodium ions have been shown to be relatively mobile under the influence of high electric fields within a silicon dioxide insulating layer, and the resulting movement of charge can result in unwanted instabilities and a drift of critical device parameters. Although the mobilities of the positive ions above in silicon dioxide are quite small, their movement through the gate insulator can be appreciable because of the extremely high electric field strengths present under typical operating conditions. For example, the application of a 10-V gate bias to a device with a gate insulator which has a thickness of 1000 Å will result in an electric field intensity of 10^6 V/cm within the insulator, neglecting the voltage impressed across the surface space-charge region. Referring to the structure shown in Figure 6.6a, both mobile and nonmobile positive ions are typically found within the silicon dioxide gate insulator of a MOS device. (Mobile ions are distinguished from immobile ions in the figure in that the latter are pictured with circles around them.) The application of a positive gate-to-substrate potential, as shown in Figure 6.6b, will result in a large electric field in the oxide in such a direction as to repel the mobile positive ions away from the gate electrode and toward the silicon-silicon dioxide interface. As these ions drift toward the interface, they will tend to induce more and more negative charge at the silicon surface. For example, as shown in the structure of Figure 6.6, this can result in an increase in the amount of surface inversion observed in a MOS device fabricated on a moderately high-resistivity *p*-type substrate. If the surface conductance is plotted as a function of time, it will rapidly increase immediately after the application of a positive gate voltage, but will eventually saturate when all the mobile ions have drifted to the silicon-silicon dioxide interface under the influence of the applied field. Even if the gate voltage is returned to zero, the surface conductance will not revert back to its original value as shown in Figure 6.7. Only the application of a negative voltage to the gate electrode can restore the surface conductance to its original value by drawing the mobile positive ions away from the interface and back into the bulk of the insulating layer.

Similarly, if mobile positive charge is located within the gate oxide, the application of positive gate voltage to a MOS capacitor fabricated on an *n*-type substrate will result in increased surface *accumulation*. In general, if the gate oxide layer of a MOS capacitor is contaminated with sodium or any other positive mobile ionic species, the application of a positive gate-to-substrate voltage will tend to shift the observed flatband voltage in the negative direction, while the application of a negative gate voltage will result in a positive shift.

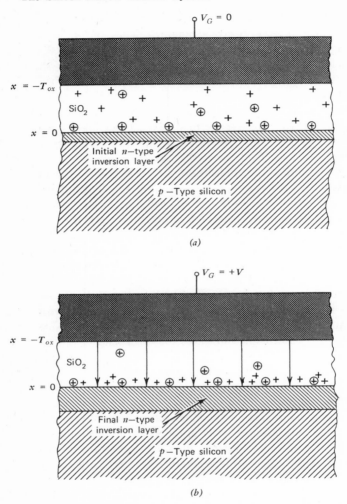

FIGURE 6.6 Drift of mobile positive ions in silicon dioxide under the influence of a strong electric field. (a) initial distribution before the application of any gate voltage. (b) final distribution after the application of positive gate voltage.

The problem of electric field induced ionic drift in the gate insulators of MOS structures can be effectively minimized in a number of ways. The most commonly employed approach is to eliminate, through the use of ultraclean processing techniques, as much sodium ion contamination in the gate insulator as possible. For example, great care must be taken to keep the quartz walls of the furnace in which the silicon dioxide gate insulator is

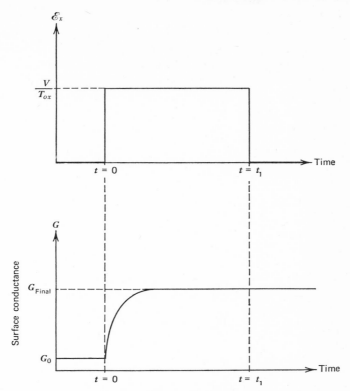

FIGURE 6.7   Electric field induced drift of surface conductance as a function of time.

grown virtually sodium-free. Recently, Kriegler, Cheng, and Colton[20] have reported that a mixture of hydrogen chloride gas and dry oxygen is extremely effective for the "cleaning" of quartz furnace tubes. They also found that the addition of a small percentage of hydrogen chloride or chlorine to the oxidizing atmosphere significantly improved the electrical stability of silicon dioxide films grown in the presence of dry oxygen. They found that this technique not only decreased the mobile ion contamination originating from the furnace tube, but also tended to passivate the oxide films grown in this manner against ionic instabilities caused by the subsequent deposition of a contaminated metalization layer.

After the growth of a silicon dioxide layer, the oxide may become contaminated through the application of an overlying layer of photo resist which might be required for the next photolithographic operation. Since most photo resists are usually heavily contaminated with sodium, the oxide layer should be chemically cleaned and etched slightly after the photo operation has been completed to remove the top 50 to 150 Å of oxide along with any residual contaminants. This step is especially critical in the case of the gate oxide, after the contact photo step, and prior to the deposition of the metallic layer which will eventually form the gate electrode. However, the cleaning operation is not necessary with silicon-gate or molybdenum-gate processing techniques in which the gate electrode is used as a mask to selectively etch the underlying silicon dioxide layer, since in these structures the photo resist never comes in direct contact with the gate insulator. (Silicon-gate and molybdenum-gate technologies are discussed in detail in the next chapter.)

As previously mentioned, the reason that mobile positive ion migration is possible through the silicon dioxide layer of a typical MOS device is because the mobility of the ionic species within the silicon dioxide is sufficiently large to result in an appreciable drift velocity at room temperature for electric fields on the order of 10^5 to 10^6 V/cm. (The mobility of the ions, hence their drift velocity for a given value of applied field, will increase with increasing temperature. However, while the ionic migration process can be accelerated at higher temperatures, the maximum value of shift in either the flatband voltage or the threshold voltage will be independent of temperature.) Other approaches to minimizing the instabilities in MOS structures caused by positive ion migration usually attempt in some way to reduce the mobility of the ionic species within the gate dielectric. One technique that has been used successfully to reduce the movement of positive ions through the gate insulator is the formation of a thin layer of phosphosilicate glass over the thermally grown silicon dioxide gate insulator.[21] Using radioactive tracers, Yon, Ko, and Kuper[22] showed that sodium ions tended to be much more soluble in the phosphosilicate glass than in the oxide below and consequently were gettered by the glass layer, thus preventing them from drifting across the gate insulator under the influence of the applied gate field. If this technique is used, however, care must be taken to ensure that the thickness of the phosphosilicate glass layer is small compared to the thickness of the remaining silicon dioxide. If this is not done, a dipole-type polarization can occur that can itself result in a shift of either the flatband or the threshold voltage, independent of the degree of sodium ion contamination within the gate insulator.[23] Although the resulting shift will be in the same direction for a given polarity gate voltage as would be observed in the case of positive ion migration, unlike the latter mechanism, the final value of the shift resulting from this polarization will be a function of the *magnitude* of the applied gate voltage.

Both silicon nitride and aluminum oxide have been shown to be extremely effective barriers to sodium ion migration in the presence of high electric fields.[24,25] To prevent sodium from entering the gate insulator of MOS structures while also maintaining the highly desired properties of the interface between silicon and its thermally grown oxide, a number of workers have fabricated devices using double-layer gate insulators. The stability and electrical characteristics of these structures have been extremely good, with large scale integrated circuits employing both SiO_2-Si_3N_4 and SiO_2-Al_2O_3 gate dielectrics now being routinely fabricated in large volume production. A further advantage of employing these double-layer gate insulators is the fact that both silicon nitride and aluminum oxide have dielectric constants which are substantially greater than the dielectric constant of silicon dioxide. Consequently, for the same value of *mechanical thickness* as an insulating layer consisting only of silicon dioxide, a double-layer insulator consisting of either SiO_2-Si_3N_4 or SiO_2-Al_2O_3 will be electrically equivalent to a *thinner* layer of silicon dioxide. This will, of course, result in a higher gain factor, β, for the device and will reduce the amount of change in the threshold voltage with substrate-to-source voltage (source-body effect). For p-channel configurations, the use of either of the above double-layer insulators can also result in a lowering of the threshold voltage to values not achievable with similar devices fabricated with silicon dioxide gate insulators of equal *mechanical* thickness, thus enabling integrated circuits fabricated through the use of these techniques to exhibit threshold voltages typically in the -1.5 to -2.3 V range and thereby allowing these circuits to interface directly with bipolar integrated circuits.

Although MOS structures fabricated with either SiO_2-Al_2O_3 or SiO_2-Si_3N_4 gate dielectrics have been shown to virtually eliminate the bias instabilities associated with the drift of positive ionic species through the gate insulator, the difference in conductivities between the insulating layers can give rise to charge accumulation at the interface between them. If charge is accumulated at this interface, it will tend to induce charge of opposite polarity at the surface of the silicon, thereby shifting the observed values of flatband and threshold voltages. For a given value of applied voltage, if there is a difference in the relative conductivities of the two materials which form the double-layer gate dielectric, more current will flow through one insulating layer than through the other and charge will continue to accumulate at the interface between the two insulators until enough charge has been trapped at the interface for an equilibrium situation to exist. Consequently, because the charge accumulation at the interface will vary with time, so will the measured flatband and threshold voltage shifts. Frohman-Bentchkowsky and Forsythe[26,27] have reported that for MOS structures fabricated with gate insulators consisting of a thermally grown layer of silicon dioxide covered by an overlying layer of chemical-vapor-deposited silicon nitride, the effects of the charge accumulation above can be minimized by employing a silicon dioxide layer which is thicker than the layer of silicon nitride and by depositing a very low-conductivity nitride. In this manner, threshold voltage shifts of a few tenths of a volt can easily be achieved during operation at 125°C with 15-V applied gate bias for over 100,000 hr.[26,27] The low-conductivity nitride can be obtained by carefully controlling the silane-to-ammonia or silane-to-silicon tetrachloride ratio during the deposition of the silicon nitride layer and also by depositing the nitride at relatively low temperatures, in the range 700 to 850°C.

Early work performed by Nigh, Stach, and Jacobs indicated that equally good stability and performance could be achieved by using devices fabricated with SiO_2-Al_2O_3 double-layer gate dielectrics.[24] This was subsequently confirmed by Cheney, Jacobs, Korb, Nigh, and Stach, who used the technique to fabricate high-performance, reliable, beam-leaded MOS integrated circuits[28] and by Dudley and Labuda who reported on their excellent stability.[29] Excellent stability in SiO_2-Al_2O_3 MOS structures has also been observed by Curry and Nigh[25] and Walden and Strain,[30] so long as moderately low operating voltages and electric field strengths within the gate insulators were maintained.

In general, the stability of double-insulator MOS device structures will improve as the ratio of the thickness of the silicon dioxide layer to the thickness of the overlying alumina or nitride layer is increased. For stable operation of digital MOS integrated circuits fabricated through the use of the oxide-nitride technology, for example, a "sandwich" consisting of 600 Å of silicon dioxide covered by a 400 Å layer of silicon nitride is commonly employed.

It is interesting to note that, for a very *thin* silicon dioxide layer on the order of 15 to 50 Å covered by a relatively thick layer of silicon nitride, appreciable charge accumulation can be realized at the interface between the two insulators as a result of charge tunneling directly through the thin oxide region and being trapped in localized states at the oxide-nitride interface. Donor-type states or traps have been shown to exist at the boundary between the silicon nitride and the underlying silicon dioxide. With the application of a relatively large negative gate voltage, the states can become positively charged as electrons are repelled out of the traps and tunnel through the very thin layer of silicon dioxide into the substrate. Similarly, with the application of a large positive voltage, the electrons will tunnel back into the traps from the silicon substrate. The amount of charge located in the traps can be varied over a wide range, and since the oxide-nitride interface is in very close proximity to the silicon surface, will have a pronounced effect on the surface potential of the silicon and, in turn, on the threshold voltage of the device. For *p*-channel structures, for example, the application of a large negative gate voltage will typically accumulate enough positive charge at the oxide-nitride interface as to make the threshold voltage of the

MOSFET *highly negative*. The subsequent application of a large positive gate voltage will shift the threshold voltage back to a low negative value and the device will once again exhibit its initial characteristics. A great deal of interest has developed in direct-tunneling metal-nitride-oxide-silicon (MNOS) structures because of their electrically-alterable threshold voltage characteristics and their applications to nonvolatile electrically-alterable semiconductor electronic memories.[31-35]

6.5   THE INFLUENCE OF THE OXIDE-SILICON INTERFACE ON THE OBSERVED CARRIER MOBILITY IN THE CHANNEL REGION OF A MOS FIELD-EFFECT TRANSISTOR

The current flowing through an insulated-gate MOSFET is directly proportional to the effective mobility of the carriers in the channel region as they travel from source to drain. In deriving the equations that are typically used to describe the electrical characteristics of MOS devices, the assumption is usually made that the mobility of the carriers in the channel is a constant, independent of the applied gate and drain voltages. However, while this approximation holds quite well to first order for typical MOSFETs, it has been demonstrated that the assumption of constant carrier mobility is no longer valid in both the case where a device is operating with large values of applied gate voltage and also in the case in which a device is fabricated with a very short channel length and is operating with large values of applied drain voltage. Furthermore, the field-effect mobility that characterizes the movement of the free carriers in the channel region is usually substantially less than the mobility observed deep in the bulk of the substrate, and has been shown to be a function of the crystallographic orientation of the semiconductor substrate and the amount of diffuse scattering experienced by the carriers at the surface of the semiconductor, as well as the factors which influence the mobility within the bulk.

The mobility of a carrier traveling through any material will be determined by the number of different types of scattering mechanisms it can experience during its movement and the frequency of these scattering events or collisions. The collisions may scatter the carrier in any direction, and it can easily be seen how, over a prolonged period of time, the successive scattering events determine an average drift velocity for the carrier.[36] In general, the average time between collisions will be given by

$$\frac{1}{\tau} = \frac{1}{\tau_1} + \frac{1}{\tau_2} + \ldots + \frac{1}{\tau_n} = \sum_{x=1}^{n} \frac{1}{\tau_x}, \qquad (6.1)$$

where τ_n is the average time between collisions associated with the nth scattering mechanism. Similarly, the carrier mobility is given by

$$\frac{1}{\mu} = \frac{1}{\mu_1} + \frac{1}{\mu_2} + \ldots + \frac{1}{\mu_n} = \sum_{x=1}^{n} \frac{1}{\mu_x}, \tag{6.2}$$

where μ_n is the mobility associated with the nth scattering mechanism. Typically, within the bulk of the semiconductor substrate, carriers moving through the material will experience collisions with both lattice and impurity sites; thus the carrier mobility will be given simply by

$$\frac{1}{\mu} = \frac{1}{\mu_{\text{lattice}}} + \frac{1}{\mu_{\text{impurity}}}. \tag{6.3}$$

However, since most of the conduction in a MOSFET occurs near the surface of the semiconductor and in close proximity to the insulator-semiconductor interface, an additional scattering mechanism is observed that tends to reduce the effective mobility of the carriers, particularly at high gate voltages.

6.5.1   Variation of Surface Mobility with Applied Gate Voltage

Typically, the gate insulator in a MOSFET is about 1000 Å thick and, for gate voltages of 10 V or higher, electric field strengths of 10^6 V/cm or more will be impressed across the insulator. When the perpendicular electric field in the insulator is so high, it represents an "irresistible" force of such magnitude that carriers traveling through the channel from source to drain are attracted to the semiconductor-insulator interface and bounce off it continually as they traverse the total distance of the gap. Ideally, if the interface is perfectly smooth, only specular scattering will occur; the y-component of vector momentum will remain unchanged after each collision and the carrier mobility will remain unaffected. However, since the interface between the semiconductor and the insulator is uneven and imperfect, diffuse or partially diffuse scattering will be observed that will tend to reduce the average value of both the carrier drift velocity in the y-direction and the y-component of vector momentum, thereby decreasing the effective carrier mobility. Consequently,

$$\frac{1}{\mu} = \frac{1}{\mu_{\text{lattice}}} + \frac{1}{\mu_{\text{impurity}}} + \frac{1}{\mu_{\text{surface}}}. \tag{6.4}$$

As the perpendicular field strength increases with increasing gate voltage, the effect of the surface scattering term in (6.4) will increase and the effective carrier mobility will decrease. The more disordered the interface, the more it will act to randomize the y-component of the carrier velocity after a collision and the lower the observed carrier mobility in the channel will be at a given gate voltage.

A theory of surface mobility on semiconductor substrates based on diffuse surface scattering was first proposed by Schrieffer[37] and was latter modified and improved by Greene, Frankl, and Zemel.[38] Fang and Triebwasser showed that, by assuming diffuse scattering of monoenergetic electrons, a theoretical expression could be derived that gave good agreement with experimental results obtained from measurements of inversion layers on *p*-type silicon.[39] While models based on diffuse and partially diffuse surface scattering have correlated well with experimental measurements of actual device parameters, some deficiencies have been pointed out, and other researchers have turned to explaining observed results with a model that assumes carrier scattering off localized surface charges at the interface. Neumark[40] has proposed a model for analysis of mobility in MOS inversion layers based on *misfit dislocations* (interfacial dislocations that have been observed to form a two-dimensional network of lines or long rods in the plane of the boundary between the semiconductor and the insulator). Neumark has indicated that the consideration of misfit dislocations is essential to any analysis of the silicon-silicon dioxide interface and, in particular, to surface mobility behavior in general. These types of dislocations are expected purely from an energetic standpoint in cases of lattice mismatch and have even been observed in systems with much less lattice-mismatch than in the silicon-silicon dioxide system (i.e., GaAs-Ge). Neumark assumes that experimentally observed surface states which act as carrier traps at the silicon dioxide-silicon interface are not randomly distributed but, instead, are states associated with the misfit dislocations and are bunched along dislocation lines. The surface states, for the case of *n*-type inversion layers on *p*-type silicon substrates, are shown to be negatively charged when the inversion layer forms, because of localized electron trapping. Thus free electrons are repelled from the surface around each dislocation as they traverse the gap from source to drain, and the resulting surface scattering mechanism results in a reduction of the carrier mobility. If the effect of the charge repulsion potential were to extend downward and entirely through the entire inversion layer, the resulting reduction in mobility would be much greater than if the dislocations were to occur within the bulk of the semi-conductor. Indeed, the misfit dislocation model yields a predicted surface mobility behavior that appears to agree quantitatively with experimental data obtained on *p*-type silicon with both (100) and (111) orientations. In fact, since there is a greater degree of lattice mismatch at the (111) silicon-silicon dioxide interface than at the (100) silicon-silicon dioxide interface, if one assumes that the misfit dislocation density increases with the degree of mismatch, the misfit dislocations may be linked to both the fixed positive charge density per unit area at the oxide-silicon interface, Q_{SS}, and to the so-called fast surface states observed at the interface, since these quantities are always found to be greater on (111) material than on (100) material.

Carrier mobilities for both p- and n-type inversion layers have also been shown to be a strong function of the orientation of the silicon substrate.[41,42] As a rule, hole mobilities associated with p-channel devices fabricated on (111) silicon substrates have been found to be greater than with similar devices fabricated on (100) substrates. However, the observed values of surface electron mobility for n-channel devices fabricated on (100) orientation p-type silicon have been found to be consistently higher than for similar devices fabricated on (111) orientation p-type material. While these results are not fully understood, Colman, Bate, and Mize[41] have attributed the dependence of the carrier mobilities on the orientation of the substrate to different effective masses, as a function of orientation, at the surface of the silicon for the quantized carriers.

6.5.2 The Effect of the Variation of Carrier Mobility with Applied Gate Voltage on the Electrical Characteristics of MOS Field-Effect Transistors

As a result of the additional surface scattering term in (6.4), the mobilities associated with carriers traveling through the channel region of a MOSFET have been found to be lower than in the bulk of the semiconductor. (Bulk values of both electron and hole mobilities in silicon at room temperature as a function of substrate doping concentration as tabulated by Phillips[43] are shown in Figure 6.8.) As discussed in the previous section, the effective mobilities of the carriers will exhibit a further reduction with increasing gate voltage. The decrease in mobility is very small at low values of V_G, but the effect increases rapidly as the magnitude of the gate voltage is increased.[44-46]

It should be remembered that the equations that were derived in Chapter 4 to describe the drain current-versus-drain voltage characteristics of MOSFETs as a function of the applied gate voltage were developed under the assumption of constant carrier mobility in the channel region. To obtain closer agreement at higher values of gate voltage between these equations and experimentally observed device characteristics, one can simply substitute an empirically-determined mobility-versus-gate voltage relationship instead of assuming a constant value.[45]

Richman[47] has reported that the effects of the variation in carrier mobility with applied gate voltage, along with the effects of any parasitic resistances that may be present in the drain or source circuits, are particularly pronounced when MOSFETs are operated at low values of applied drain-to-source voltage, well below pinch-off. Referring to Figure 6.9, which illustrates

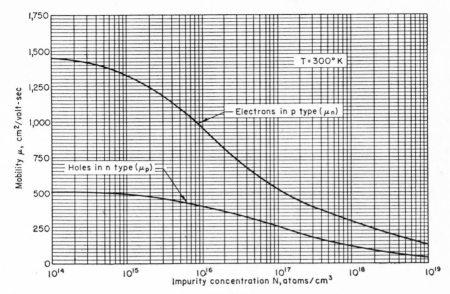

FIGURE 6.8   Minority carrier drift mobilities in bulk silicon at room temperature as a function of impurity doping concentration. (After Phillips[43].)

the drain current-versus-drain voltage characteristics associated with a typical n-channel MOSFET as a function of the applied gate voltage, if the drain currents I_{D_2} and I_{D_1} are measured at two different values of applied gate voltage, V_{G_2} and V_{G_1}, respectively, at a constant value of drain voltage, then it follows that for V_D sufficiently small that the device is operating well below pinch-off,

$$I_{D_2} \cong \beta V_D[(V_{G_2} - V_T) - \tfrac{1}{2}V_D], \qquad (6.5)$$

and

$$I_{D_1} \cong \beta V_D[(V_{G_1} - V_T) - \tfrac{1}{2}V_D]. \qquad (6.6)$$

Subtraction of (6.6) from (6.5) yields

$$I_{D_2} - I_{D_1} \cong \beta V_D[V_{G_2} - V_{G_1}]. \qquad (6.7)$$

It should be noted that the result would be exactly the same if the more exact form for the drain current below pinch-off, as given by (4.14), were used in (6.5) and (6.6).

The V_{G_2} and V_{G_1} (hence I_{D_2} and I_{D_1}) were chosen arbitrarily. Consequently, (6.7) predicts that for a given *change in the gate voltage, there will be a*

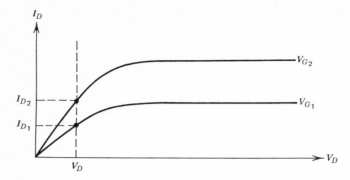

FIGURE 6.9 Current intercepts at constant drain-to-source voltage for an *n*-channel MOSFET operating below pinch-off. (After Richman[47].)

corresponding change in the drain current, independent of the initial value of the gate voltage. In other words, (6.7) predicts that, for a fixed value of drain voltage, the transconductance of the device will be a constant at all current levels when the device is operating below pinch-off. However, it will be seen that this prediction breaks down at large values of gate voltage.

Equation 6.7 predicts that, for a typical MOSFET, if the drain current increases by 100 μA when the gate voltage minus the threshold voltage is increased from 1 to 2 V at a constant drain voltage well below pinch-off, the drain current will again increase by 100 μA when the gate voltage minus the threshold voltage is raised, for example, from 9 to 10 V. In other words, as the gate voltage is increased, the drain current will keep increasing at the same rate, as illustrated in Figure 6.10. Thus for very high gate voltages (yet lower than the breakdown voltage associated with the gate dielectric layer) the resistance between the drain and source for a typical MOSFET operating below pinch-off might approach zero. This cannot happen for two reasons. First, as previously discussed, the mobility associated with the carriers in the channel region will begin to degrade with increasing gate voltage, and as a result, the transconductance of the device will decrease as the gate voltage is increased. Second, a minimum series resistance will always be observed between drain and source because of the parasitic resistances within the diffused drain and source regions, the metallic contacts to these regions, and the wire interconnections between the bonding pads and the package leads. Consequently, because of gate-field-dependent mobility and the effects of parasitic series resistances in the drain and source circuits, one can expect to find that the drain-to-source conductance of a MOSFET that is operating well below pinch-off will approach a limiting value when the device is

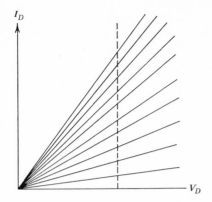

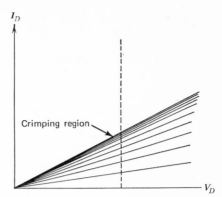

FIGURE 6.10 "Ideal" characteristics associated with an n-channel MOSFET operating below pinch-off. (After Richman[47].)

FIGURE 6.11 Effect of crimping on the characteristics of an n-channel MOSFET operating below pinch-off. (After Richman[47].)

operated at very high values of applied gate voltage, and the current-versus-voltage characteristics will tend to "crimp" under these conditions, as is illustrated in Figure 6.11. Although which of the two mechanisms discussed, gate-field-dependent mobility or series parasitic resistance, is dominant will vary according to the individual characteristics of different MOS devices, the simple resistive model shown in Figure 6.12 can be used to accurately

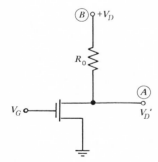

FIGURE 6.12 Model for a MOS transistor with limiting series drain resistance. (After Richman[47].)

describe the effects of either or both mechanisms on the electrical characteristics of a device operating below pinch-off.[47] As is illustrated in the figure, a resistor of value R_0 is placed in series with the drain electrode of an ideal MOS transistor, that is a device whose i-v characteristics before pinch-off behave according to (4.14). If the device is fabricated on a moderately high-resistivity p-type silicon substrate and the assumption is made that the effects of the charge in the surface depletion region are relatively small, then at point A the drain current associated with the n-channel MOSFET is approximately given by

$$I_D \cong \beta V_{D'}[(V_G - V_T) - \tfrac{1}{2}V_{D'}]. \tag{6.8}$$

In the model, the actual drain electrode of the nonideal transistor is at point B. The drain current I_D also flows through the series resistor R_0 such that

$$V_{D'} = V_D - I_D R_0. \tag{6.9}$$

Substituting (6.9) into (6.8) yields

$$I_D \cong \beta(V_D - I_D R_0)[(V_G - V_T) - \tfrac{1}{2}(V_D - I_D R_0)]. \tag{6.10}$$

Assuming that the term $I_D R_0$ is relatively small compared to the drain voltage V_D and, consequently, that $(I_D R_0)^2$ is negligible compared to other terms, (6.10) reduces to

$$I_D \cong \beta V_D[(V_G - V_T) - \tfrac{1}{2}V_D] + \beta V_D I_D R_0 - \beta I_D R_0(V_G - V_T). \tag{6.11}$$

Therefore,

$$I_D[1 + \beta R_0(V_G - V_T - V_D)] \cong \beta V_D[(V_G - V_T - \tfrac{1}{2}V_D)] \tag{6.12}$$

or

$$I_D \cong \frac{\beta V_D[(V_G - V_T) - \tfrac{1}{2}V_D]}{1 + \beta R_0[(V_G - V_T) - V_D]}. \tag{6.13}$$

Equation 6.13 can be used to describe the current-versus-voltage characteristics of an n-channel MOSFET operating with relatively large applied gate voltage. (A similar expression for p-channel devices can easily be derived in the same way.) The actual value of R_0 can be determined by simply measuring the behavior of the device in question under these conditions. It can easily be seen that (6.13) gives the correct dependence, since in the limit of applied gate voltages sufficiently low that no crimping occurs,

$$\beta R_0[(V_G - V_T) - V_D] \ll 1 \tag{6.14}$$

and (6.13) reduces to the form of (4.14) (remembering that the effects of the charge in the surface depletion region have been assumed to be negligible). On the other hand, for gate voltages sufficiently high that appreciable crimping is observed,

$$\beta R_0[(V_G - V_T) - V_D] \gg 1 \tag{6.15}$$

and

$$(V_G - V_T) \gg V_D > \tfrac{1}{2}V_D. \tag{6.16}$$

Under these conditions, the second term in the denominator of (6.13) becomes dominant, and the drain current approaches the limiting case imposed by the presence of the series resistor R_0. That is, for large values of the gate voltage minus the threshold voltage,

$$I_D \cong \frac{\beta V_D(V_G - V_T)}{\beta R_0(V_G - V_T)} = \frac{V_D}{R_0}. \tag{6.17}$$

It can be seen that the ideal MOS transistor is essentially a short circuit under these conditions and the drain-to-source current-versus-voltage characteristics of the device below pinch-off are dominated by the effective series resistor.

Both the model proposed by Richman to describe the effects of gate-field-dependent mobility and parasitic series resistances, and the resulting behavior predicted by (6.13) have been shown by Lunsmann to exhibit extremely good agreement with experimental data obtained from measurements of typical MOS devices.[48]

As was the case when a MOSFET is operated at relatively low values of drain voltage well below pinch-off, the effects of mobility degradation at high gate voltages and of parasitic resistances in the drain and source circuits can have an appreciable effect on the electrical characteristics of a device that is operating in the saturated drain current region. In particular, these two mechanisms will tend to limit the range over which the square-law dependence of the drain current beyond pinch-off on the square of the gate voltage minus the threshold voltage, as given by (4.41), is valid.

6.5.3 Variation of Surface Mobility with Applied Drain Voltage

Although first-order theory predicts that the transconductance of a MOSFET is inversely proportional to the spacing between the diffused drain and source regions, for very small channel lengths this dependence no longer holds and, in fact, the transconductance per unit channel width of a silicon device approaches a limiting value, because at very high values of transverse electric field, the mobility of the carriers in the channel is no longer a constant at a given gate voltage but begins to decrease with increasing *transverse* field strength. Beyond a critical value of electric field, saturation of carrier drift velocity occurs and the carrier mobility is observed to vary inversely with electric field. This phenomenon is extremely important in predicting the electrical characteristics of MOS field-effect devices fabricated with very short channel lengths, on the order of a few microns or less. The characteristics of short-channel MOSFETs operating under velocity-saturated conditions are described in greater detail in Chapter 7.

REFERENCES

1. E. H. Nicollian and A. Goetzberger, The Si-SiO$_2$ Interface Electrical Properties as Determined by the Metal-Insulator-Silicon Conductance Technique, *Bell System Technical Journal*, Vol. 46, 1967, pp. 1055–1133.

2. P. V. Gray and D. M. Brown, Density of Si-SiO$_2$ Interface States, *Applied Physics Letters*, Vol. 8, 1966, pp. 31–33.

3. D. M. Brown and P. V. Gray, Si-SiO$_2$ Fast Interface State Measurements, *Journal of the Electrochemical Society*, Vol. 115, 1968, pp. 760–766.

4. P. L. Castro and B. E. Deal, Low Temperature Reduction of Fast Surface States Associated with Thermally Oxidized Silicon, *Journal of Electrochemical Society*, Vol. 118, 1971, pp. 280–286.

5. B. E. Deal, M. Sklar, A. S. Grove, and E. H. Snow, Characteristics of the Surface State Charge (Q_{SS}) of Thermally Oxidized Silicon, *Journal of the Electrochemical Society*, Vol. 114, 1967, pp. 226–274.

6. F. Faggin and T. Klein, Silicon Gate Technology, *Solid State Electronics*, Vol. 13, 1970, pp. 1125–1144.

7. P. V. Gray, The Silicon-Silicon Dioxide System, *Proceedings of the IEEE*, Vol. 57, No. 9, September 1969, pp. 1543–1551.

8. A. G. Revesz, K. H. Zaininger, and R. J. Evans, Interface States and Interface Disorder in the Si-SiO$_2$ System, *Journal of Physics and Chemistry of Solids*, Vol. 28, 1967, pp. 197–204.

9. S. R. Hofstein, Stabilization of MOS Devices, *Solid State Electronics*, Vol. 10, 1967, pp. 657–670.

10. M. M. Atalla and E. Tannenbaum, Impurity Redistribution and Junction Formation in Silicon by Thermal Oxidation, *Bell System Technical Journal*, Vol. 39, 1960, p. 933.

11. A. S. Grove, O. Leistiko, and C. T. Sah, Redistribution of Acceptor and Donor Impurities During Thermal Oxidation of Silicon, *Journal of Applied Physics*, Vol. 35, 1964, p. 2695.

12. B. E. Deal, A. S. Grove, E. H. Snow, and C. T. Sah, Observation of Impurity Redistribution During Thermal Oxidation of Silicon Using the MOS Structure, *Journal of the Electrochemical Society*, Vol. 112, 1965, pp. 308–314.

13. A. S. Grove, *Physics and Technology of Semiconductor Devices*, John Wiley and Sons, New York, 1967, pp. 69–74.

14. G. Schottky, Decrease of F.E.T. Threshold Voltage due to Boron Depletion During Thermal Oxidation, *Solid State Electronics*, Vol. 14, 1971, pp. 467–474.

15. K. H. Zaininger, Electron Bombardment of MOS Capacitors, *Applied Physics Letters*, Vol. 8, 1966, pp. 140–142.

16. A. S. Grove and E. H. Snow, A Model for Radiation Damage in Metal Oxide Semiconductor Structures, *Proceedings of the IEEE*, Vol. 54, 1966, pp. 894–895.

17. K. H. Zaininger and A. S. Waxman, Radiation Resistance of Al$_2$O$_3$ MOS Devices, *IEEE Transactions on Electron Devices*, Vol. ED-16, No. 4, 1969, pp. 333–338.

18. B. Andre, J. Buxo, D. Esteve, and H. Martinot, Effects of Ionizing Radiation on MOS Devices, *Solid State Electronics*, Vol. 12, 1969, pp. 123–131.

19. R. P. Donovan and M. Simons, Radiation Hardening of Thermal Oxides on Silicon

via Ion Implantation, presented at the 1969 Government Microcircuits Applications Conference, Washington, D.C.

20. R. J. Kriegler, Y. C. Cheng, and D. R. Colton, The Effect of HCl and Cl_2 on the Thermal Oxidation of Silicon, *Journal of the Electrochemical Society*, Vol. 119, No. 3, March 1972, pp. 388–392.

21. D. R. Kerr, J. S. Logan, P. J. Burkhardt, and W. A. Pliskin, Stabilization of SiO_2 Passivation Layers with P_2O_5, *IBM Journal of Research and Development*, Vol. 8, 1964, p. 376.

22. E. Yon, W. H. Ko, and A. B. Kuper, Sodium Distribution in Thermal Oxide by Radiochemical and MOS analysis, *IEEE Transactions on Electron Devices*, Vol. ED-13, 1966, p. 276.

23. E. H. Snow and B. E. Deal, Polarization Phenomena and Other Properties of Phosphosilicate Glass Films on Silicon, *Journal of the Electrochemical Society*, Vol. 113, 1966, p. 263.

24. H. E. Nigh, J. Stach, and R. M. Jacobs, A Sealed-Gate IGFET, presented at the 1967 Solid State Device Research Conference, Santa Barbara, California, June 1967.

25. J. J. Curry and H. E. Nigh, Bias Instability in Two-Layer MIS Structures, presented at the 1970 IEEE Reliability Physics Symposium, Las Vegas, Nevada, April 1970.

26. D. Frohman-Bentchkowsky and D. D. Forsythe, Reliability of MNOS Integrated Circuits, presented at the 1969 International Electron Device Meeting, Washington, D.C., October 1969.

27. D. Frohman-Bentchkowsky, The Metal-Nitride-Oxide-Silicon (MNOS) Transistor-Characteristics and Applications, *Proceedings of the IEEE*, Vol. 58, No. 8, August 1970, pp. 1207–1219.

28. G. T. Cheney, R. M. Jacobs, H. W. Korb, H. E. Nigh, and J. Stach, Al_2O_3–SiO_2 IGFET Integrated Circuits, presented at the 1967 International Electron Device Meeting, Washington, D.C., October 1967.

29. R. H. Dudley and E. F. Labuda, A. Reliable Contact for Insulated Gate Field Effect Integrated Circuits, presented at the 1969 International Electron Device Meeting, Washington, D.C., October 1969.

30. R. H. Walden and R. J. Strain, Conduction in Films of Pyrolytic Al_2O_3, presented at the 1970 IEEE Reliability Physics Symposium, Las Vegas, Nevada, April 1970.

31. J. T. Wallmark and J. H. Scott, Switching and Storage Characteristics of MIS Memory Transistors, *RCA Review*, Vol. 30, No. 2, June 1969, pp. 335–365.

32. E. C. Ross, M. T. Duffy, and A. M. Goodman, Effects of Silicon Nitride Growth Temperature on Charge Storage in the MNOS Structure, *Applied Physics Letters*, Vol. 15, No. 12, December 15 1969, pp. 408–409.

33. E. C. Ross, A. M. Goodman, and M. T. Duffy, Operational Dependence of the Direct-Tunneling Mode MNOS Memory Transistor on the SiO_2 Layer Thickness, *RCA Review*, Vol. 31, No. 3, September 1970, pp. 467–478.

34. R. F. Vieth, Nitride-Oxide Layer Proofs Memory Against Data Loss, *Electronics*, July 5, 1971, pp. 53–56.

35. C. Svensson, Theory of the Maximum Charge Stored in the Thin Oxide MNOS Memory Transistor, *Proceedings of the IEEE*, Vol 59, No. 7, July 1971, pp. 1134–1136.

36. P. Richman, *Characteristics and Operation of MOS Field-Effect Devices*, McGraw-Hill Book Co., New York, 1967, pp. 10–13.

37. J. R. Schrieffer, Effective Carrier Mobility in Surface Space Charge Layers, *Physical Review*, Vol. 97, 1955, pp. 641–646.

38. R. F. Greene, D. R. Frankl, and J. Zemel, Surface Transport in Semiconductors, *Physical Review*, Vol. 118, 1960, pp. 967–975.

39. F. Fang and S. Triebwasser, Effect of Surface Scattering on Electron Mobility in an Inversion Layer on *p*-type Silicon, *Applied Physics Letters*, Vol. 4, No. 8, April 1964, pp. 145–147.

40. G. F. Neumark, New Model for Interface Charge-Carrier Mobility: The Role of Misfit Dislocations, *Physical Review Letters*, Vol. 21, No. 17, October 1968, pp. 1252–1256.

41. D. Coleman, R. T. Bate, and J. P. Mize, Mobility Anisotropy and Piezoresistance in Silicon *p*-type Inversion Layers, *Journal of Applied Physics*, Vol. 39, No. 4, March 1968, pp. 1923–1931.

42. N. St. J. Murphy, F. Berz, and I. Flinn, Carrier Mobility in Silicon MOSTs, *Solid State Electronics*, Vol. 12, 1969, pp. 775–786

43. A. B. Phillips, *Transistor Engineering*, McGraw-Hill Book Co., New York, 1962, pp. 67–71.

44. O. Leistiko, Jr., A. S. Grove, and C. T. Sah, Electron and Hole Mobilities in Inversion Layers on Thermally Oxidized Silicon Surfaces, *IEEE Transactions on Electron Devices*, Vol. ED-12, No. 5, May 1965, pp. 248–254.

45. D. Frohman-Bentchkowsky, On the Effect of Mobility Variation on MOS Device Characteristics, *Proceedings of the IEEE*, Vol. 56, No. 2, February 1968, pp. 217–218.

46. V. G. K. Reddi, Majority Carrier Surface Mobilities in Thermally Oxidized Silicon, *IEEE Transactions on Electron Devices*, Vol. ED-15, No. 3, March 1968, pp. 151–160.

47. P. Richman, *Characteristics and Operation of MOS Field-Effect Devices*, McGraw-Hill Book Co., New York, 1967, pp. 50–56.

48. P. D. Lunsmann, Equivalent Circuit for the Conductance of a MOSFET, *Electronics Letters*, Vol. 4, No. 6, March 1968, pp. 100–101.

BIBLIOGRAPHY

Bandali, M. B., The Effects of the Field Dependence of Carrier Mobility on the Validity of the Gradual Channel Approximation in Insulated-Gate Field-Effect Transistors, *Solid State Electronics*, Vol. 14, 1971, pp. 1325–1327.

Bar-Lev, A., and S Margalit, Changes of Mobility Along a Depletion Type MOS Transistor Channel, *Solid State Electronics*, Vol. 13, 1970, pp. 1541–1546.

Berz, F., Ionized Impurity Scattering in Silicon Surface Channels, *Solid State Electronics*, Vol. 13, 1970, pp. 903–906.

Deal, B. E., E. L. MacKenna, and P. L. Castro, Characteristics of Fast Surface States Associated with SiO_2-Si and Si_3N_4-SiO_2-Si Structures, *Journal of the Electrochemical Society*, Vol. 116, No. 7, July 1969, pp. 997–1005.

Deuling, H., E. Klausmann, and A. Goetzberger, Interface States in Si-SiO_2 Interfaces, *Solid State Electronics*, Vol. 15, 1972, pp. 559–571.

Eversteyn, F. C., and H. L. Peek, Preparation and Stability of Enhancement *n*-channel MOS Transistors with High Electron Mobility, *Philips Research Reports*, Vol. 24, 1969, pp. 15–33.

Goetzberger, A., V. Heine, and E. H. Nicollian, Surface States in Silicon from Charges in the Oxide Coating, *Applied Physics Letters*, Vol. 12, No. 3, February 1968, pp. 95–97.

Gregor, L. V., Passivation of Semiconductor Surfaces, *Solid State Technology*, April 1971, pp. 37–43.

Gwyn, C. W., Model for Radiation-Induced Charge Trapping and Annealing in the Oxide Layer of MOS Devices, *Journal of Applied Physics*, Vol. 40, No. 12, November 1969, pp. 4886–4892.

Holmes-Siedle, A. G., and K. H. Zaininger, Designing MOS Systems for Radiation Environments, *Solid State Technology*, May 1969, pp. 40–49, 71.

Jund, C., B. Kervella, and J. Grosvalet, A Mechanism of Evaluation in MAOS Systems, presented at the 1971 IEEE Reliability Physics Symposium, Las Vegas, Nevada, March-April 1971.

Kar, S., and W. E. Dahlke, Interface States in MOS Structures with 20–40 Å Thick SiO_2 Films on Non-Degenerate Silicon, *Solid State Electronics*, Vol. 15, 1972, pp. 221–237.

Koomen, J., The Measurement of Interface State Charge in the MOS System, *Solid State Electronics*, Vol. 14, 1971, pp. 571–580.

Lundstrom, K. I., and C. M. Svensson, Properties of MNOS Structures, *IEEE Transactions on Electron Devices*, Vol. ED-19, No. 6, June 1972, pp. 826–836.

Mansour, I. R. M., E. A. Talkhan, and A. I. Barboor, Investigations on the Effect of Drift-Field-Dependent Mobility on MOST Characteristics, *IEEE Transactions on Electron Devices*, Vol. ED-19, No. 8, August 1972, pp. 899–916.

Margalit, S., A. Neugroschel, and A. Bar-Lev, Redistribution of Boron and Phosphorus in Silicon After Two Oxidation Steps Used in MOST Fabrication, *IEEE Transactions on Electron Devices*, Vol. ED-19, No. 7, July 1972, pp. 861–868.

Newman, P. A., and H. A. R. Wegener, Effect of Electron Irradiation on Silicon Nitride Insulated Gate Field Effect Transistors, *IEEE Transactions on Nuclear Science*, Vol. NS-14, No. 6, December 1967, pp. 293–298.

Rodriguez, V., and M. A. Nicolet, Drift Velocity of Electrons in Silicon at High Electric Fields from 4.2° to 300°K., *Journal of Applied Physics*, Vol. 40, No. 2, February 1969, pp. 496–498.

Sandor, J. E., A Model for the "Fixed" Charge at the Si/SiO_2 Interface, *Proceedings of the IEEE*, Vol. 57, No. 6, June 1969, pp. 1184–1186.

Severi, M., and G. Soncini, Surface State Density at the (Hydrogen-Chloride) Oxide-Silicon Interface, *Electronics Letters*, Vol. 8, No. 16, August 1972, pp. 402–404.

Snow, E. H., A. S. Grove, B. E. Deal, and C. T. Sah, Ion Transport Phenomena in Insulating Films, *Journal of Applied Physics*, Vol. 36, No. 5, May 1965, pp. 1664–1673.

Stanley, A. G., Effect of Electron Irradiation on Carrier Mobilities in Inversion Layers of Insulated Gate Field Effect Transistors, *IEEE Transactions on Nuclear Science*, Vol. NS-14, No. 6, December 1967, pp. 266–275.

Svensson, C., and I. Lundstrom, Theory of the Thin-Oxide MNOS Memory Transistor, *Electronics Letters*, Vol. 6, No. 20, October 1970, pp. 645–647.

Whelan, M. V., On the Nature of Interface States in an SiO_2-Si System, and on the Influence of Heat Treatments on Oxide Charge, *Philips Research Reports*, Vol. 22, 1967, pp. 289–303.

PROBLEMS

6.1 Although p-channel MOS integrated circuits do not usually operate with an applied substrate voltage, the source-body effect will influence the characteristics of certain off-ground devices such as load resistors. Discuss the effects of impurity redistribution during thermal oxidation on the electrical characteristics of these devices, in particular on the source-body effect.

6.2 In the fabrication of thick-oxide aluminum-gate MOS integrated circuits, the diffusion of the drain, source, and silicon interconnection regions is immediately followed by a prolonged high-temperature wet oxidation to form the thick oxide. In p-channel configurations, this usually results in a substantial increase in the sheet resistivity of the diffused areas compared to the value measured before oxidation. Discuss the role of impurity redistribution during thermal oxidation in this effect. Will the same behavior be observed in the fabrication of n-channel structures? What will happen? Why?

6.3 The sheet resistivity associated with the diffused regions of p-channel silicon-gate MOS integrated circuits is typically much lower than the diffused sheet resistivity of p-channel thick-oxide aluminum-gate MOS structures. Why? (For a discussion of typical silicon-gate processing techniques, see Chapter 7.)

6.4 Discuss the effects of substrate resistivity, if any, on the degree of degradation of the electrical characteristics of MOS devices which have been bombarded by ionizing radiation. In particular, discuss (*a*) space-charge build-up and the resulting shifts in flatband and threshold voltages, and (*b*) the formation of interface states at the oxide-silicon interface.

6.5 Why might n-channel digital integrated circuits typically be more susceptible to the effects of ionizing radiation than similar p-channel circuits?

6.6 Consider a MOS capacitor with a silicon dioxide gate insulator of thickness T_{ox} fabricated on a silicon substrate. Initially, a uniform density per unit volume, N, of positively charged sodium ions exists within the gate insulator between the metal gate electrode at $x = -T_{ox}$ and the silicon-silicon dioxide interface at $x = 0$. At time $t = 0$, a voltage of $+V_0$ volts is applied to the gate electrode. Assuming that all the electric field resulting from the application of the gate voltage exists totally within the gate insulator, derive an expression for the change in the flatband voltage of the device as a function of time. Assume μ_{Na^+} is the mobility of sodium ions in silicon dioxide.

6.7 Referring to problem 6.6, after a sufficient amount of time, all the sodium ions will have accumulated at the silicon-silicon dioxide interface. Now, at time $t = t_1$, a voltage of $-V_0$ volts is applied to the gate electrode. What is the expression for the change in flatband voltage as a function of time for all t greater than t_1? Does the charge distribution in the insulator ever return to the initial condition of problem 6.6? Why?

6.8 Consider two MOS capacitors formed on a silicon substrate. The first capacitor consists of 600 Å of silicon dioxide covered by a 400-Å layer of silicon nitride and an overlying metal electrode. The second capacitor consists of an 800-Å layer of silicon dioxide with an overlying metal electrode. Both capacitors have been contaminated with a layer of 10^{11} sodium ions per square centimeter in the silicon dioxide, in close proximity to the oxide-silicon interface. A large negative voltage is applied to both gate electrodes. Assuming that the dielectric constant of silicon nitride is twice that of silicon dioxide, what will the final shift in the flatband voltage observed for each capacitor be?

6.9 Besides being a very effective barrier to sodium ion migration, silicon nitride is also a very effective moisture and oxidation barrier. Consider a p-channel MOS transistor structure fabricated with a double-layer gate dielectric consisting of 600 Å of silicon dioxide covered by 400 Å of silicon nitride. The silicon nitride was deposited at 800°C and selectively etched using a subsequently deposited layer of pyrolytic silicon dioxide as an etching mask. Prior to the photolithographic operation, the deposited oxide was densified at 850°C in wet oxygen for 15 min, and then in dry oxygen for an additional 15 min. Why will the observed value of Q_{SS} for this structure always be at the minimum value associated with the orientation of the silicon substrate? Why will it be impossible to achieve a high threshold voltage and, simultaneously, a high gain factor β with this structure?

6.10 Derive an expression for the drain current of an n-channel MOSFET operating well below pinch-off as a function of the applied gate voltage if the (ideal) device has a series resistance R_{00} in the source circuit as well as a series resistance R_0 in the drain circuit. Discuss your result.

6.11 Derive an expression for the saturated drain current beyond pinch-off as a function of the gate voltage for the n-channel device described in the previous problem. Discuss your result. Ideally, does the resistor in the drain circuit have any effect?

7

High-Frequency Operation
of MOS Field-Effect Devices

7.1 SPEED LIMITATIONS OF CONVENTIONAL MOS FIELD-EFFECT TRANSISTORS AND INTEGRATED CIRCUITS

In general, silicon bipolar transistors are capable of operating at much higher frequencies when compared to conventional silicon MOSFETs. One of the major reasons for the relatively slow speed of the MOSFET is the presence of an appreciable gate-to-drain capacitance. For an enhancement-type device, since there is no initial channel present with zero gate-to-source voltage, the gate electrode must be required to extend completely over the gap region between the drain and source so that a conducting channel can be formed between the two regions when the appropriate polarity gate voltage is applied. The gate electrode is usually designed to overlap both the drain and source electrodes so that any slight misalignment in the positioning of the gate electrode during the fabrication process will not result in a device failure. Thus because of the overlap of the gate and drain and the gate and source electrodes and because of the thin insulating layer that separates them, appreciable values of parasitic gate-to-drain and gate-to-source capacitances will be observed in conventional MOSFET structures. In particular, since the output of a MOS amplifier (or inverter) stage is taken from the drain electrode while the input is applied to the gate electrode, the gate-to-drain capacitance can be considered as a feedback capacitor from the output back to the input. Since the output is 180 degrees out of phase with respect to the input signal because of the inversion in the amplifier, the feedback signal through the gate-to-drain capacitance is *negative feedback*. As the frequency of operation

173

is increased, the effect of the negative feedback capacitance also increases and the observed gain begins to decrease rapidly.

Another very important consideration in any comparison of the relative speeds of bipolar and MOS devices is the difference in typical values of transconductance per unit area between the two types of structures. Because the transconductance per unit area of a silicon MOSFET fabricated through the use of conventional processing techniques is typically less than that of a silicon bipolar transistor operating at a comparable current level, the time required by the MOSFET to either charge or discharge a given capacitive load will be greater than the time required by the bipolar transistor. The gate input impedance of a MOSFET is, for all practical purposes, purely capacitive, and this capacitance must be charged or discharged to change the state of the device. Therefore, in integrated circuit applications where the gate input capacitance associated with a particular stage must be charged and discharged by the previous stage, the longer times required by MOS devices for charging and discharging will result in slower circuit speeds.

The ratio of the transconductance to the gate input capacitance of a MOSFET can be used as a figure of merit for the speed of the device or for the relative speed of integrated circuits employing interconnected combinations of similar devices. The (g_m/C_{in}) ratio can be increased by the following:

1. Driving the MOSFET harder by applying higher values of gate-to-source voltage.

2. Increasing the mobility of the carriers in the inversion layer.

3. Decreasing the drain-to-source spacing.

4. Decreasing the gate input capacitance by decreasing the parasitic capacitances associated with the amount of overlap of the gate electrode over the drain and source regions.

The transconductance of a MOSFET can, of course, be increased by increasing the channel width of the device. However, the gate input capacitance will also increase proportionally and the (g_m/C_{in}) ratio will remain unchanged. Similarly, the use of a material with a higher dielectric constant for the gate insulator or decreasing the thickness of the gate insulator will have no effect on this ratio, since both the transconductance and the gate input capacitance will increase by the same factor. (When the effect of a constant value of stray capacitance is considered, however, a second-order speed increase in speed can result when these quantities are increased.)

7.2 AN EQUIVALENT CIRCUIT FOR MOS FIELD-EFFECT TRANSISTORS

By examining the structure of the MOSFET, the physical properties of the device can be related to a generalized equivalent circuit consisting of discrete

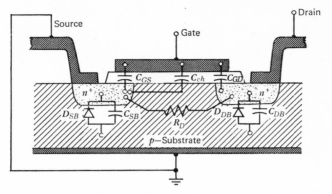

FIGURE 7.1   Cross-sectional representation of an *n*-channel MOSFET with associated discrete components to be used in equivalent circuit model.

components. This equivalent circuit, in turn, can be used to predict the performance of MOSFETs as a function of frequency.[1] A cross-sectional view of a typical (*n*-channel) MOSFET is shown in Figure 7.1. The individual discrete components that can be physically associated with each part of the structure are also shown in the figure, superimposed in the appropriate regions. The symbols used for each are defined below:

C_{ch} = the capacitance from the gate electrode to the active channel area.

C_{DB} = the drain-to-substrate capacitance across the junction as a function of the drain-to-substrate potential.

C_{SB} = the source-to-substrate capacitance across the junction as a function of the source-to-substrate potential.

C_{GD} = the gate-to-drain negative feedback capacitance.

C_{GS} = the gate-to-source capacitance.

D_{DB} = the diode formed between the diffused drain region and the substrate.

D_{SB} = the diode formed between the diffused source region and the substrate.

R_D = the dynamic drain-to-source resistance.

The total drain current flowing through the MOSFET can be expressed as

$$I_D = I_{D0} + i, \tag{7.1}$$

where I_{D0} denotes the steady-state value of the drain current at a particular fixed d.c. operating point and i denotes any *incremental* change in the drain current. Similarly, the applied gate voltage can be expressed as

$$V_G = V_{G0} + v_G, \tag{7.2}$$

where V_{G0} denotes the steady-state value of the gate voltage at the d.c. operating point and v_G denotes any incremental change in gate voltage. The incremental change in the drain current as a function of any incremental change in the gate voltage with the drain voltage held constant can be written as

$$i = dI_D = \left(\frac{dI_D}{dV_G}\right)_{V_D} dV_G = \left(\frac{dI_D}{dV_G}\right)_{V_D} v_G. \tag{7.3}$$

Since the transconductance is defined as

$$g_m = \left(\frac{dI_D}{dV_G}\right)_{V_D}, \tag{7.4}$$

it follows that the incremental change in the drain current as a function of the incremental change in the applied gate voltage will be simply

$$i = g_m v_G. \tag{7.5}$$

Under normal operating conditions, the source and the substrate are usually connected in common; therefore, diode D_{SB} will be short-circuited along with capacitor C_{SB}. Furthermore, since the applied drain voltage will be positive for an n-channel device, diode D_{DB} will be open-circuited and can also effectively be removed from the equivalent circuit. The capacitance of the reverse-biased junction (C_{DB}) will, of course, vary with the drain-to-substrate voltage. It should be pointed out that the directions of diodes D_{DB} and D_{SB} will be reversed with respect to the directions shown in Figure 7.1 for the case of a p-channel MOSFET. However, for a properly biased p-channel device with a common source and substrate, both diodes can again be removed from the equivalent circuit, which will then be equally

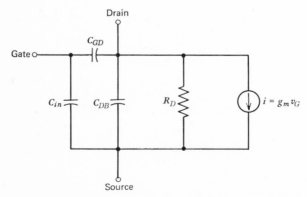

FIGURE 7.2   A small-signal equivalent circuit for an n-channel MOSFET.

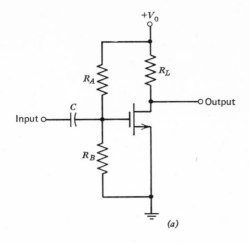

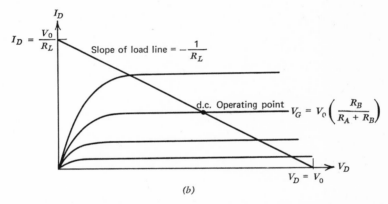

FIGURE 7.3 (*a*) Linear MOS amplifier circuit, (*b*) three-terminal device characteristics with superimposed load line. (After Richman[1].)

valid for both *p*- and *n*-channel MOSFETs. This incremental equivalent circuit is shown in Figure 7.2, where the input capacitance from gate-to-source, C_{in}, is equal to the parallel combination of C_{GS} and C_{ch}.

The incremental equivalent circuit shown in the figure can now be used to examine the behavior of any MOSFET circuit as a function of frequency. For example, consider the simple linear amplifier shown in Figure 7.3, which consists of an *n*-channel MOSFET with a resistive load, R_L. The *C* is a blocking capacitor, and resistors R_A and R_B are so chosen that the transistor is biased well within the region of saturated drain current. If the circuit is driven with a small-signal voltage source of magnitude v_s having an internal

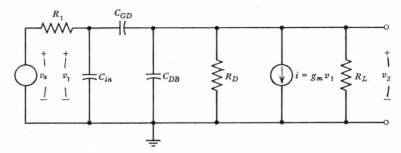

FIGURE 7.4 Equivalent circuit for the linear MOS amplifier of Figure 7.3.

resistance R_1, the equivalent incremental circuit for the amplifier will be as illustrated in Figure 7.4. By applying Norton's theorem and substituting an equivalent small-signal input current source, the equivalent circuit for the linear amplifier can be described completely in terms of admittances as shown in the simplified form of Figure 7.5. The individual admittances are given by

$$Y_1 = \frac{1}{R_1} + j\omega C_{in}, \tag{7.6}$$

$$Y_2 = j\omega C_{GD}, \tag{7.7}$$

and

$$Y_3 = \frac{1}{R_D} + j\omega C_{DB}. \tag{7.8}$$

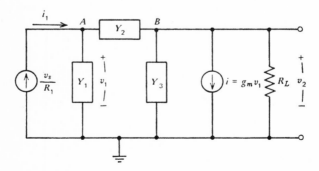

FIGURE 7.5 Equivalent circuit for MOS linear amplifier in terms of admittances. (After Richman[1].)

By applying Kirchhoff's current law to nodes A and B, one obtains at node A

$$Y_1 v_1 + Y_2(v_1 - v_2) = \frac{v_s}{R_1} \tag{7.9}$$

and at node B

$$Y_2(v_2 - v_1) + g_m v_1 + \left(Y_3 + \frac{1}{R_L}\right) v_2 = 0. \tag{7.10}$$

Equations 7.9 and 7.10 can be expressed in matrix form as

$$\begin{vmatrix} (Y_1 + Y_2) & (-Y_2) \\ (-Y_2 + g_m) & \left(+Y_2 + Y_3 + \dfrac{1}{R_L}\right) \end{vmatrix} \begin{vmatrix} v_1 \\ v_2 \end{vmatrix} = \begin{vmatrix} \dfrac{v_s}{R_1} \\ 0 \end{vmatrix}. \tag{7.11}$$

Now, defining

$$A \equiv \begin{vmatrix} (Y_1 + Y_2) & (-Y_2) \\ (-Y_2 + g_m) & \left(+Y_2 + Y_3 + \dfrac{1}{R_L}\right) \end{vmatrix}, \tag{7.12}$$

the determinant of the matrix A is

$$\det A = (Y_1 + Y_2)\left(Y_2 + Y_3 + \frac{1}{R_L}\right) + Y_2(-Y_2 + g_m). \tag{7.13}$$

The inverse of A is therefore given by

$$A^{-1} = \frac{1}{\det A} \begin{vmatrix} \left(+Y_2 + Y_3 + \dfrac{1}{R_L}\right) & (+Y_2) \\ (+Y_2 - g_m) & (+Y_1 + Y_2) \end{vmatrix}. \tag{7.14}$$

Since $AA^{-1} = I$, where I is the identity matrix, it follows that (7.11) can be multiplied by A^{-1} to yield

$$A^{-1}A \begin{vmatrix} v_1 \\ v_2 \end{vmatrix} = \begin{vmatrix} v_1 \\ v_2 \end{vmatrix} = A^{-1} \begin{vmatrix} \dfrac{v_s}{R_1} \\ 0 \end{vmatrix}. \tag{7.15}$$

Substituting (7.14) into (7.15) and performing the necessary matrix multiplication gives

$$\begin{vmatrix} v_1 \\ v_2 \end{vmatrix} = \frac{1}{\det A} \begin{vmatrix} \left(Y_2 + Y_3 + \dfrac{1}{R_L}\right)\dfrac{v_s}{R_1} \\ (Y_2 - g_m)\dfrac{v_s}{R_1} \end{vmatrix}. \tag{7.16}$$

Therefore, the small-signal output voltage will be given by

$$v_2 = \frac{v_s}{\det A}\left(\frac{Y_2 - g_m}{R_1}\right),$$ (7.17)

and the incremental gain of the amplifier is

$$G \equiv \frac{v_2}{v_s} = \frac{1}{\det A}\left(\frac{Y_2 - g_m}{R_1}\right).$$ (7.18)

Substituting (7.13) into (7.18), multiplying, and rearranging terms yields the following expression for the incremental amplifier gain:

$$G = \frac{Y_2 - g_m}{R_1\{Y_1 Y_2 + Y_1 Y_3 + Y_2 Y_3 + [(Y_1 + Y_2)/R_L] + g_m Y_2\}}.$$ (7.19)

For low-frequency operation, the capacitive admittances will be comparatively small, and consequently, for small values of ω, Y_1, Y_2, and Y_3 can be approximated by

$$Y_1 \simeq \frac{1}{R_1}, \qquad Y_2 \simeq 0, \qquad Y_3 \simeq \frac{1}{R_D}.$$ (7.20)

Thus at low frequencies, the generalized expression for the incremental gain of the linear MOS amplifier is approximately given by

$$G \simeq \frac{-g_m R_L R_D}{R_L + R_D}.$$ (7.21)

Equation 7.21 can be rewritten in terms of the amplification factor, $\mu \equiv g_m R_D$, as

$$G \simeq \frac{-\mu R_L}{R_L + R_D}.$$ (7.22)

If the MOS transistor exhibits a relatively large drain saturation resistance when operated beyond pinch-off (i.e. $R_D \gg R_L$) the low-frequency gain will be approximately equal to $-g_m R_L$. At higher frequencies, the approximations stated in (7.20), (7.21), and (7.22) will no longer be valid, the effects of the capacitive admittances will become appreciable, and the gain of the amplifier will decrease. Under these conditions, (7.19) must be used to calculate the amplifier gain.

The high-frequency performance of the amplifier, and of MOS integrated circuits in general, can be greatly improved by reducing the parasitic capacitances associated with the MOSFET structure. Of course, the active channel capacitance C_{ch} cannot be reduced without a proportional decrease in the

of the high-frequency gain of the MOS amplifier can be approximately described in terms of the low-frequency gain G_0 by the simple voltage-divider circuit shown in Figure 7.6.

7.3   TRANSIT TIME CONSIDERATIONS

For a perfectly self-aligned IGFET structure fabricated on a lightly doped substrate in which the effects of the parasitic gate-to-drain, gate-to-source, and drain-to-substrate capacitances are negligible, the maximum frequency of operation of the device will be limited by the transit time of the carriers

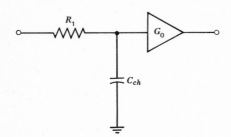

FIGURE 7.6 Simple voltage-divider circuit for relating the high-frequency gain of a linear MOS amplifier to its low-frequency gain, for an (ideal) MOSFET with a self-aligned gate electrode which is fabricated on a lightly doped substrate.

as they travel from source to drain. Under these conditions, the maximum operating frequency will be

$$\omega_{max} = 2\pi f_{max} = \frac{1}{t}, \qquad (7.29)$$

where the transit time t is given by

$$t = \frac{L}{\mu\,|\mathscr{E}|} = \frac{L^2}{\mu\,|V|}. \qquad (7.30)$$

Now, V will be equal to the drain voltage V_D when the transistor is operated below pinch-off. However, when the device is operated beyond pinch-off in the region of saturated drain current, $V_G - V_T$ will be impressed across the channel and $V_D - (V_G - V_T)$ will be impressed across the drain depletion region. If the width of the drain depletion region is much smaller than the length of the channel region, the time required for the carriers to cross the depletion region will be small compared to the time required for them to traverse the channel because of the relatively high electric field intensity that is present near the drain. Consequently, the source-to-drain transit time for drain voltages beyond pinch-off will approximately be equal to

$$t \simeq \frac{L^2}{\mu\,|V_G - V_T|}, \qquad (7.31)$$

and the maximum frequency of operation for the device can be expressed as

$$\omega_{\max} = 2\pi f_{\max} \cong \frac{\mu |V_G - V_T|}{L^2} = \frac{\beta |V_G - V_T|}{C_{ch}} = \frac{g_m}{C_{ch}}. \qquad (7.32)$$

A similar expression can easily be derived for drain voltages below pinch-off. (See the problems at the end of this chapter.)

At very small values of drain-to-source spacing, the mobility of the carriers may become a function of the applied drain voltage and (7.30), (7.31), and (7.32) should be modified accordingly.

7.4 HIGH-FREQUENCY INSULATED-GATE FIELD-EFFECT TRANSISTOR STRUCTURES

Recently, a considerable amount of research has been done toward the development of IGFET device structures that are capable of operating at higher frequencies than the conventional *p*-channel silicon MOSFET. Since the maximum frequency of operation of an insulated-gate field-effect transistor can be increased by increasing the ratio of the transconductance to the active channel capacitance and eliminating all parasitic capacitances associated with the device structure, these efforts have generally been in three areas:

1. Attempts to employ high-mobility substrate materials for IGFET structures.
2. Attempts to minimize parasitic gate-to-drain and gate-to-source capacitances in devices with small channel lengths through the use of IGFET structures with self-aligned gate electrodes.
3. Attempts to solve the problems associated with fabricating IGFET structures with extremely small channel lengths to achieve both high transconductance and low active channel capacitance.

Recent developments in all of the areas above are treated in detail in the following sections, and other approaches to fabricating high-speed MOSFET devices are also discussed.

7.5 MOSFETS FABRICATED ON HIGH-MOBILITY MATERIALS

7.5.1 *n*-Channel Silicon MOSFETs and Integrated Circuits

As discussed in Section 7.3, the maximum frequency of operation associated with a MOS device can be shown to be directly proportional to the mobility of the carriers in the active channel region. Thus since it is a well-known fact that the mobility of electrons in silicon at room temperature is considerably

n-channel MOS integrated circuits fabricated on moderately high-resistivity p-type silicon substrates and operating with negative substrate bias have been shown to function at considerably higher speeds than comparable p-channel circuits.[2]

Probably the most significant problem encountered in the fabrication of early n-channel integrated circuits was that of *field inversion*. As discussed in Section 2.2.1, it is desirable to have the thick-field threshold voltage (or *field-inversion voltage*) associated with a metal interconnection passing over a thick oxide layer which crosses two nonrelated diffused regions to be *as high as possible*, so that the active devices may be driven with high voltage levels without any possibility of undesired parasitic conducting paths forming between neighboring transistors or diffused regions. This condition can easily be achieved on a routine basis for p-channel integrated circuit structures, but is not easily realized in n-channel configurations. Early n-channel integrated circuits were found to be particularly susceptible to failures resulting from cross-coupling between neighboring nonrelated active diffused regions because of inherently low field-inversion voltages and because of instability mechanisms that tended to even further reduce the initially observed values of field-inversion voltage as a function of time under normal operating conditions. These problems occurred because the total positive ionic charge density (including both mobile and nonmobile components) in the thick oxide regions of MOS structures generally tends to be substantially greater than in the active thin gate oxide regions. Thus while this phenomenon actually *increases* the ratio of the thick-field threshold voltage to the active threshold voltage for p-channel MOS integrated circuits, it will have the reverse effect on n-channel structures and the ratio of the thick-field threshold voltage to the active threshold voltage will *decrease*, thereby limiting the voltage with which the n-channel configuration may be driven. Consequently, while it is possible to routinely fabricate p-channel MOS integrated circuits that are characterized by active device threshold voltages of approximately -2.0 V and field-inversion voltages in excess of -20 V, it was initially extremely difficult to obtain equally good results (i.e., $V_T \cong +2.0$ V and $V_{TT} > +20$ V) on a reproducible basis with early n-channel fabrication techniques.

Since the *mobile* component of the positive ionic charge density in the thick oxide regions of a MOS integrated circuit structure is also usually much greater than in the active thin gate oxide regions, the field-inversion voltage will be particularly susceptible to long-term drift caused by the migration of the positive mobile ions in the thick oxide regions under the influence of an applied electric field. In the case of an n-channel MOS integrated circuit, for which the applied voltage on metal interconnections is almost always positive, the positive voltage will repel any mobile positive ionic charge in the thick oxide underlying the metal and drive it toward the oxide-silicon interface, where it will attract an equal and opposite amount of negative charge, thereby reducing the field-inversion voltage below its initial value.

A number of different fabrication techniques have been developed recently to eliminate the problems discussed above and thereby make possible the realization of reliable high-speed n-channel MOS integrated circuits. Perhaps the most attractive of these techniques is the n-channel *Coplamos* process developed by Richman and Hayes.[3] This technology, which is discussed in greater detail in a later section, relies on selectively increasing the surface acceptor doping concentration directly under all thick oxide regions within the integrated circuit structure. In this way, it is possible to raise the field-inversion voltage in Coplamos structures to levels far greater than even in conventional p-channel integrated circuits. Referring to (2.44) and (2.80), *both the threshold voltage associated with a MOS device and the change in its threshold voltage as a function of an applied substrate-to-source potential are linearly proportional to the quantity* $Q_{SD_{max}}T_{ox}$. If the *effective acceptor doping concentration* in the thick-field regions of an n-channel MOS integrated circuit structure is increased *selectively* by doping only the portions of the silicon surface directly beneath the thick oxide, it follows that the maximum value of the charge density per unit area in the surface depletion region at these locations will become substantially greater than in the active areas of the circuit. Furthermore, since the thick oxide regions in typical MOS integrated circuits are usually on the order of 15 to 20 times thicker than the thin oxide layers that form the gate insulators of active devices, it is apparent that the magnitude of the quantity $Q_{SD_{max}}T_{ox}$ associated with the thick oxide regions can be made to be extremely large compared to the magnitude of the same quantity in the thin oxide regions. Consequently, it can easily be seen that by using such a technique it is possible to increase the field-inversion voltage with the applied substrate-to-source voltage set equal to zero and, with the application of a negative substrate voltage, increase its value to levels far in excess of those typically observed in conventional n- or p-channel structures. Richman and Hayes have reported that they were able to achieve active threshold voltages typically on the order of $+1.7$ V with a fixed -5 V applied substrate voltage while maintaining thick-field threshold voltages typically in excess of $+70$ V. At zero substrate bias, the active threshold and thick-field threshold voltages were typically $+0.3$ and $+15$ V, respectively.[3]

The selective doping of the surface of the silicon underlying the thick oxide regions also virtually eliminates the tendency of parasitic space-charge-limited currents to flow between closely spaced nonrelated diffused regions. This is accomplished because the increased acceptor concentration at the surface of the silicon sharply decreases the lateral spreading of the depletion

grown on low-resistivity cadmium-doped p-type GaAs substrates. Once again, the drain and source regions were tin-doped. With the use of silicon nitride gate insulators, a substantial decrease in the interface state charge density was also achieved. The high-frequency performance of these gallium arsenide devices was far superior when compared to similar MOSFETs fabricated on silicon substrates. The gallium arsenide transistors were characterized by electron field-effect mobilities as high as 2400 cm²/V-sec and, as a result, were able to achieve power gains on the order of 22 dB at 200 MHz—approximately 3 to 6 dB higher than those of silicon devices. As expected, because of the wide energy band gap, the gallium arsenide MOSFETs performed very well at high temperatures, exhibiting power gains at 200 MHz as high as 9 dB at 300°C. In contrast, the silicon devices exhibited 0 dB power gain at 260°C.

7.6 REDUCTION OF PARASITIC INTERELECTRODE CAPACITANCES IN IGFET STRUCTURES

7.6.1 The Offset-Gate Depletion-Type MOSFET

Although the gate electrode is required to extend completely over the gap region between the drain and the source for an *enhancement-type* device, this is not necessarily so for a *depletion-type* MOSFET. If a conducting surface channel is present between the drain and source, the flow of current between the two regions can be modulated by an offset insulated-gate structure as shown in Figure 7.7 for an n-channel MOSFET. By offsetting the gate electrode away from the drain, the negative feedback capacitance C_{GD} can be substantially reduced at the expense of the introduction of a small series drain resistance that results from the unmodulated portion of the channel near the drain region. Although pinch-off will be observed at slightly larger values of applied drain voltage, very little degradation in device gain will result from the parasitic drain resistance as long as its magnitude is much less than typical values of the dynamic drain saturation resistance. On the other hand, if the gate electrode is offset away from the *source region*, serious *source degeneration* will occur and the observed gain of the transistor will decrease substantially.

Offset-gate n-channel silicon MOSFETs similar to the structure shown in Figure 7.7 have been fabricated with relatively low values of series drain resistance and, because of the extremely small gate-to-drain capacitances associated with the devices, have exhibited better high-frequency characteristics when compared with similar devices in which the gate electrode overlapped both the source and drain regions. Offset-gate MOSFETs are usually of the n-channel type because of the higher mobility of electrons in silicon,

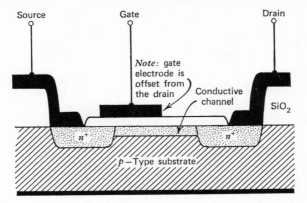

FIGURE 7.7   Cross-sectional representation of an offset-gate n-channel depletion-type MOSFET.

which greatly improves both the transconductance per unit area and the resulting high-frequency performance, and because of the relative ease with which n-channel depletion-type silicon devices can be fabricated.

It should be noted that because the thin oxide directly above the un-modulated portion of the channel in the offset-gate structure is *not* covered by metal, devices of this type are especially susceptible to electrical instabilities which can result from ionic conduction processes that may take place on the surface of the unprotected oxide layer.[19] If proper precautions are taken during device fabrication, these instabilities can be minimized, although not entirely eliminated. Another approach in reducing the negative feedback capacitance between the drain and gate electrodes in a depletion-type MOSFET is to have the gate electrode overlap both the drain and source regions while defining the region in which the thin gate insulator is to be grown so that it extends *only* over the source region. Thus the oxide over the portion of the channel that is nearer the drain diffusion is much thicker than it is near the source, and the parasitic gate-to-drain capacitance is substantially reduced. Although the reduction in parasitic capacitance is not as great as in the offset-gate structure, this type of device will not be susceptible to electrical instabilities resulting from surface ionic conduction because the oxide overlying the channel region will be completely covered by metal at all locations.

7.6.2   The Dual-Gate MOS Tetrode

The appreciable negative feedback capacitance between gate and drain that is inherent in the conventional MOSFET can, for all practical purposes, be eliminated through the use of the dual-gate MOS tetrode structure shown in

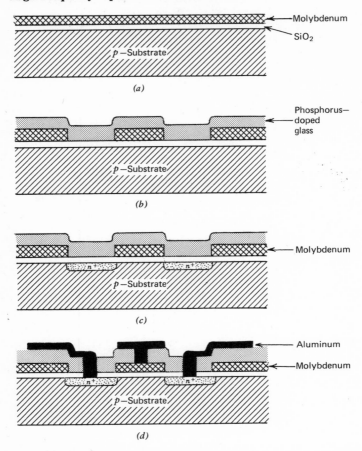

Figure 7.9 Processing sequence for the fabrication of *n*-channel molybdenum-gate MOSFETs.

an *n*-channel device, a thin insulating layer of silicon dioxide is first grown over the surface of a *p*-type silicon wafer. Next, a layer of molybdenum is deposited over the silicon dioxide and then covered with photo resist. The resist is exposed and developed and the resulting pattern is defined by etching the molybdenum in a potassium ferricyanide etch. A phosphorus-doped glass is then deposited over the entire wafer and the drain and source regions are subsequently formed as the phosphorus diffuses through the thin thermally grown oxide and into the silicon when the wafer is placed in a high-temperature ambient. The molybdenum gate acts as a diffusion barrier, and the channel region is not doped. Finally, contact windows are selectively etched through the phosphorus glass layer, aluminum is evaporated over the surface of the wafer, and the electrode metalization pattern is defined and etched using standard photo resist techniques. The completed self-aligned structure shown in Figure 7.9d is characterized by extremely small interelectrode overlaps and, consequently, minimal parasitic capacitances. In practice, the gate electrode will *slightly* overlap both the drain and source regions because of the lateral spreading of the junctions during the diffusion. Typical values of gate-drain and gate-source overlap resulting from the effects of lateral diffusion in the molybdenum-gate MOSFET are on the order of the junction depth of the drain and source regions, usually about 1 μ. On the other hand, typical values of gate-drain and gate-source overlap for conventionally processed devices are usually about 4 μ. Since the gate electrode in the molybdenum-gate MOSFET is automatically self-registered with the channel region, narrow-gate devices may be constructed without any critical alignment steps; therefore, the fabrication procedure is greatly simplified. The p-channel molybdenum-gate MOSFETs can be fabricated in the same manner as described above by using an n-type silicon substrate and employing boron-doped glass as a diffusion source. Also, by modifying a number of the previously discussed procedures, it is possible to fabricate self-aligned molybdenum-gate integrated circuits.[24,25] The processing techniques commonly employed to do this are quite similar to those used in the construction of *silicon-gate* integrated circuits, which are discussed in the following section.

7.6.5   Self-Aligned Silicon-Gate MOS Field-Effect Transistors and MOS Integrated Circuits

Deposited polycrystalline silicon material may also be used to form the gate electrode of a self-aligned MOSFET using a fabrication procedure very similar to the technique employed in the construction of the molybdenum-gate MOSFET. A film of polycrystalline silicon that is deposited over a thin insulating layer of silicon dioxide will act as an effective barrier for the selective diffusion of either boron or phosphorus impurities into an underlying silicon substrate. At the same time, the deposited polycrystalline silicon electrode will also be heavily doped and will become sufficiently conductive so that no overlying metalization will be required except for contact purposes. During this diffusion, the drain and source regions are formed in the areas where the polycrystalline silicon and the underlying silicon dioxide have been etched away, while the gate electrode acts as a mask to prevent the doping of the channel.

The fabrication of self-aligned insulated-gate field-effect transistors employing polycrystalline silicon gate electrodes was first reported by Bower and Dill,[26] and later by Sarace, Kerwin, Klein, and Edwards.[27] Shortly thereafter, Faggin, Klein, and Vadasz showed that the use of these devices

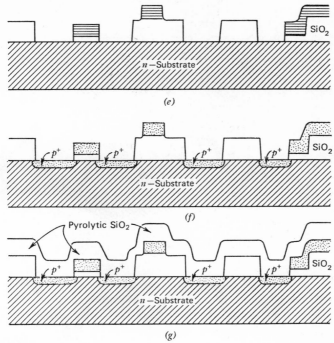

FIGURE 7.10 (*Contd.*)

The processing sequence commonly employed in the fabrication of *p*-channel silicon-gate MOS integrated circuits is illustrated in Figure 7.10. First, a relatively thick layer of silicon dioxide is thermally grown over the surface of an *n*-type silicon wafer with (111) orientation. Using standard photo resist techniques, all drain, source, channel, and p^+-interconnection regions are then simultaneously defined, and the silicon dioxide is removed from these areas through the use of a buffered hydrofluoric acid etch. Next, a thin insulating layer of silicon dioxide, which will eventually form the gate dielectric for the individual transistors, is thermally grown, and the wafer is immediately covered with a layer of pyrolytically deposited polycrystalline silicon, typically about 5000 Å thick. The wafer is then heat-treated in a dry oxygen ambient at a sufficiently high temperature so that a thin layer of silicon dioxide is formed at the surface of the deposited polycrystalline silicon layer. Photo resist is once again applied over the surface of the wafer, and the polycrystalline silicon gate and interconnection regions are defined and etched out. This is accomplished by first etching through the thin layer of silicon dioxide with buffered hydrofluoric acid and then etching through the underlying silicon layer with a mixture of nitric, hydrofluoric, and acetic

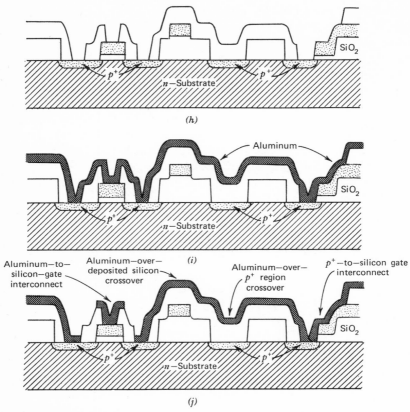

FIGURE 7.10   (*Contd.*)

acids. (In general, the adherence of the photo resist to the thin layer of silicon dioxide is much better than the adherence of the resist directly to the poly-crystalline silicon layer itself.) After the undesired regions of polycrystalline silicon have been etched away, the remaining photo resist is removed and the exposed silicon dioxide overlying the regions which will eventually form the drain, source, and p^+-interconnection regions is etched away using the polycrystalline silicon and thick oxide regions for masking purposes. Once again, buffered hydrofluoric acid is used as the etch and, in the process, the thin layer of silicon dioxide over the polycrystalline silicon regions is completely removed. At the same time, a small fraction of any thick oxide that is not covered by a polycrystalline silicon interconnection is also attacked by the etch. At this point, the wafer is placed in a diffusion furnace and all polycrystalline silicon gates and interconnections, as well as all drain, source, and p^+ interconnection regions, are heavily doped with boron. During this step, both the polycrystalline silicon regions and the thick oxide regions serve as diffusion barriers to selectively define the positions of the drains, sources, and p^+ interconnections. Since the overlying silicon-gate electrode itself acts as a diffusion mask to prevent the doping of the active channel region, the drain and source regions of each individual transistor are automatically self-registered to the gate electrode and extremely small gate-to-drain and gate-to-source overlaps will result. Because of the effects of lateral diffusion, typical values of interelectrode overlap for silicon-gate MOSFETs will be similar to values observed for molybdenum-gate devices. After the diffusion has been performed, any boron glass that has formed over the heavily doped areas is removed, and a relatively thick layer of silicon dioxide on the order of 4000 to 8000 Å is then pyrolytically deposited over the surface of the wafer. Care must be taken during the deposition of the silicon dioxide layer to ensure that the contour of the oxide as it passes over a polycrystalline silicon region is relatively smooth and free from sharp "cliffs" or "steps." If this is not accomplished, the final metalization pattern that will eventually cross over some of these steep cliffs will be very susceptible to open-circuit failures resulting from breaking, cracking, or over-etching. However, either by carefully controlling the oxide deposition conditions,[29] or by lightly doping the oxide with phosphorus as it is deposited and then *reflowing* it over the sharp steps associated with the polycrystalline silicon regions through the use of a subsequent high-temperature heat treatment,[30] a smooth contour may be achieved that will ensure the continuity of an overlying metal interconnection. After the deposited oxide is densified, the wafer is once again covered with resist, and contact windows are defined and etched through the oxide. Using electron-beam-evaporation techniques, a 15,000 to 20,000-Å layer of aluminum is then deposited over the wafer, the wafer is coated with resist, and the metal interconnection pattern is defined and selectively etched out. Finally, the wafer is treated in hydrogen at a temperature between 450 and 500°C to anneal out any radiation damage that might have occurred during the aluminum evaporation, to alloy the silicon-to-aluminum contacts, and as discussed in Section 6.1.1, to annihilate any fast surface states that might be present at the oxide-silicon interface. The completed structure, which is shown in Figure 7.10*j*, not only illustrates the self-aligned silicon-gate MOSFET, but also shows the three types of interconnections between active regions of the integrated circuit structure that are made possible through the use of silicon-gate technology: (*a*) p^+ polycrystalline silicon interconnections, (*b*) p^+ silicon interconnections within the single crystal silicon material, and (*c*) aluminum interconnections. The required cross-connections between the three interconnection levels are also illustrated, along with each necessary crossover structure.

It should be noted that any cross-connection within the integrated circuit between a polycrystalline silicon region and an adjacent p^+ diffused region at the surface of the silicon must be achieved by opening a contact hole through the deposited oxide that simultaneously exposes both regions, so that the aluminum that is subsequently evaporated will completely fill the contact hole, thus connecting the polycrystalline silicon to the diffused region. In other words, the aluminum will act as a "bridge" to connect the two regions. However, by slightly altering the fabrication sequence already described, it is possible to achieve a direct polycrystalline silicon-to-diffused region contact at the expense of an additional photolithographic operation. Previously, the polycrystalline silicon layer was deposited directly after the growth of the thin gate oxide layer. If, instead, portions of this thin oxide layer are selectively etched away (as is illustrated in Figure 7.11) at all locations where a direct contact is required, when the polycrystalline silicon is deposited, it will come into intimate contact with the silicon substrate at these locations, and the boron diffusion that follows will result in the ohmic connection shown in Figure 7.11d. As can be seen in the figure, the actual contact will be a direct result of the lateral movement of the surface diffused region and the fact that the boron will tend to diffuse all the way through the polycrystalline silicon layer and will penetrate into the single crystal silicon directly below in the areas where the thin oxide layer has been selectively etched away.

One of the major advantages of p-channel silicon-gate MOS integrated circuits is that they can be made to interface directly with bipolar transistor integrated circuits. This is possible because the threshold voltage of a p-channel silicon-gate MOSFET is always considerably lower than the threshold voltage of an aluminum-gate device when all other process and material parameters are the same. Conventional p-channel aluminum-gate MOS transistors with relatively thick silicon-dioxide gate insulators on the order of 1200 to 1500 Å that are fabricated on (111) orientation silicon substrates typically exhibit threshold voltages greater than -3 V because, as previously discussed, the effects of the fixed positive interface charge density per unit area, Q_{SS}, the charge density contained in the surface depletion region per unit area when it is at its maximum value, $Q_{SD_{max}}$, and the work function term $(\phi_{MS'} + 2\phi_F)$ associated with the MOS structure all tend to make the threshold voltage of a p-channel transistor more negative. Aluminum-gate devices fabricated on substrates with (100) orientations have been observed to exhibit lower threshold voltages because of the inherently lower values of Q_{SS} which result from the orientation of the silicon but, for the same reason, the threshold voltages associated with parasitic thick-field transistors formed by aluminum interconnections that cross over thick oxide between two adjacent p^+ regions are also lowered, thereby limiting the maximum operating voltage for the circuit. Through the use of silicon-gate technology, however, p-channel MOSFETs with typical threshold voltages between -1.5 and

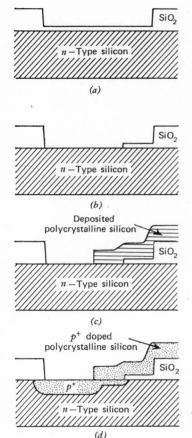

(a)

(b)

Deposited
polycrystalline silicon

SiO₂

(c)

p^+ doped
polycrystalline silicon

(d)

FIGURE 7.11 Modification in silicon-gate fabrication sequence required to achieve a direct p^+ polycrystalline silicon-to-p^+ silicon diffused region contact. The additional photolithographic operation is performed as shown in (b).

—2.5 V can easily be fabricated on (111) orientation silicon substrates so that bipolar compatibility can be achieved with no sacrifice in the maximum operating voltage of the circuit.

The p-channel silicon-gate MOSFETs exhibit lower threshold voltages than similarly constructed aluminum-gate devices because the work function term ϕ_{MS}, is *positive* for the silicon-gate structure and is *negative* for an aluminum-gate device. The reason that ϕ_{MS}, is negative for aluminum-gate MOSFETs is that the work function of aluminum is less than the work function associated with the n-type silicon substrate. However, if the material used to form the gate electrode is polycrystalline silicon that is heavily doped with boron, the work function associated with the gate electrode will be considerably *greater* than the work function of the n-type silicon substrate

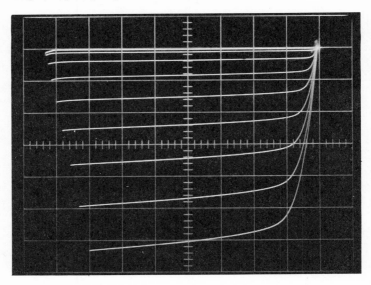

FIGURE 7.12 Electrical characteristics of a p-channel silicon-gate MOSFET fabricated on a 15 Ω-cm (100) orientation silicon substrate with $T_{ox} = 1500$ Å of silicon dioxide, $L = 1$ mil, and $W = 50$ mils; vertical scale: drain current: -1 mA/div.; horizontal scale: drain voltage: -5 V/div.; gate voltage: 0 to -8 V, in -1 V steps; substrate voltage: 0 V.

since the Fermi level in the p^+ silicon gate will lie below the valence band edge while the Fermi level in the substrate will lie above the intrinsic level within the forbidden energy gap. The net result is that, in the p-channel silicon-gate structure, the effect of the positive work function term $\phi_{MS'}$ will subtract from the effects of the fixed positive interface charge density and the maximum value of the charge density contained in the surface depletion region, thus lowering the magnitude of the threshold voltage. This is illustrated in Figures 7.12 and 7.13, which show the electrical characteristics of a p-channel silicon-gate MOSFET that was fabricated on a (100) orientation substrate with a relatively low value of Q_{SS}, which exhibited a threshold voltage exactly equal to 0 V. The device was constructed on an n-type silicon substrate with a resistivity of approximately 15 Ω-cm. Also, p-channel depletion-type devices can be fabricated by employing silicon-gate MOSFET structures on (100) orientation substrates with even higher resistivity if Q_{SS} is kept sufficiently low.

Although $\phi_{MS'}$ will be positive for a p-channel silicon-gate MOSFET, it will be negative for an n-channel silicon-gate device in which the polycrystalline silicon electrode is heavily phosphorus-doped. Since the Fermi level

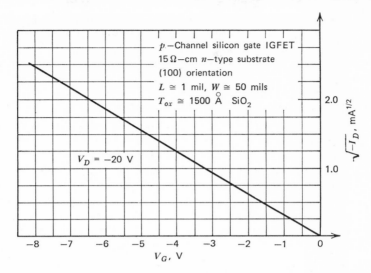

FIGURE 7.13   Plot of the square root of the negative drain current versus the applied gate voltage with constant drain voltage for the device of Figure 7.12. The extrapolated threshold voltage is approximately equal to 0 V.

in the n^+ silicon gate electrode will lie above the conduction band edge and the Fermi level in the p-type silicon substrate will lie below the intrinsic level in the forbidden energy gap, it follows that the work function associated with the substrate will be greater than the work function associated with the gate and, consequently, ϕ_{MS}, will be negative. Thus since conventional aluminum-gate n-channel MOSFETs are usually of the *depletion* type unless fabricated on low-resistivity substrates, this will also be true of silicon-gate devices. Faggin and Klein[31] found the threshold voltage of n-channel silicon-gate MOSFETs that were fabricated with heavily phosphorus-doped gate electrodes to be approximately a tenth of a volt more positive than similarly constructed aluminum-gate devices.

   In addition to the obvious advantages of higher circuit speeds resulting from the reduction of parasitic gate-to-drain and gate-to-source capacitances, and low threshold voltages for bipolar circuit compatibility, p-channel silicon-gate MOS integrated circuits have shown themselves to be superior to conventional aluminum-gate structures in a number of areas. Because the silicon-gate MOS device is self-registering, the most critical alignment step normally required to fabricate narrow-channel MOSFETs (the positioning of the gate electrode over the active channel region) is eliminated. Furthermore, because the silicon gate electrode is deposited immediately after the gate oxide is grown and covers the oxide throughout the entire processing sequence, it serves to protect the oxide from any contaminants and impurities that otherwise might penetrate it and that might eventually result in device instability. Also, the diffused p^+ regions within the silicon substrate in a silicon-gate device will exhibit substantially lower values of sheet resistivity than are commonly found in aluminum-gate devices, because the overlying thick oxide is pyrolytically deposited, rather than being thermally grown. Consequently, the boron will not be segregated into the oxide. Finally, silicon-gate MOSFETs can be made smaller than conventional devices with similar electrical characteristics because of the elimination of the additional diffused area required in a non-self-aligned structure to ensure that the gate electrode overlaps both the source and drain regions.

While silicon-gate technology can be employed to fabricate high-speed *p*-channel integrated circuits, it is equally applicable to the construction of *n*-channel ICs, with the added advantage of even higher-speed operation resulting from the increased carrier mobility. However, the processing techniques that are commonly used in the fabrication of *p*-channel silicon-gate MOS devices must be substantially modified to achieve *n*-channel structures with a sufficiently high ratio of field-inversion voltage to active device threshold voltage. As was previously discussed in Section 7.5.1, this can be accomplished through the use of the *n*-channel *Coplamos* process.[3] This technique makes it possible to diffuse selectively only the portions of the silicon surface that lie directly below the thick oxide regions so that the surface acceptor concentration in these regions becomes approximately an order of magnitude greater than the acceptor concentration in the silicon substrate itself. More important, this is achieved without the need for an additional photolithographic operation and the resulting selectively diffused regions are self-aligned to the overlying thick oxide regions. The technology is equally applicable to the construction of both *n*-channel aluminum-gate and *n*-channel silicon-gate integrated circuit structures, but only the latter configuration is considered here.

The *n*-channel *Coplamos* process makes use of the ability of an overlying layer of silicon nitride to mask against both the thermal oxidation of and the diffusion of impurities into the surface of a silicon substrate. *Coplamos* is an acronym for coplanar metal-oxide-silicon and is a fundamental extension of the early work on localized oxidation of silicon reported by Morandi,[32] and Appels, Kooi, Paffen, Schatorje, and Verkuylen,[33] who found, respectively, that a selectively etched pattern formed from either a single layer of silicon nitride or a double layer consisting of silicon dioxide covered by silicon nitride could be used to confine the thermal oxidation of a silicon wafer to specific locations where the dielectric masking layer had been etched away.

When silicon is placed in an oxygen-rich ambient at extremely high temperatures, a chemical reaction takes place between the oxygen and the silicon, and the surface begins to oxidize. Thus as the silicon dioxide is formed, it follows that the silicon surface is consumed and the oxide-silicon interface moves deeper into the substrate as the thickness of the grown oxide increases. With respect to the original silicon surface, approximately only slightly more than half of the thermally grown silicon dioxide will lie above the original surface, and the remaining portion will lie below. If, prior to thermally oxidizing a silicon wafer, a layer of silicon nitride is deposited over its surface and a pattern is defined and etched into it, the locations where the nitride remains will be masked from the effects of thermal oxidation, and the oxidation process will only take place in the areas where the surface of the silicon is exposed to the oxidizing ambient. As is illustrated in Figures 7.14a and 7.14b, approximately half of the volume of the oxide that is grown in these regions will lie below the surface of the silicon regions that are protected by the nitride. Morandi[32] showed that if at this point the thermally grown silicon dioxide regions are etched away with a buffered hydrofluoric acid etch which will not attack either the nitride or the underlying silicon substrate, and the wafer is then once again placed in the same oxidizing ambient at the same temperature for approximately the same amount of time as before, the silicon dioxide that is formed in the surrounding regions will grow such that its top surface will reach the same level as the original surface of the silicon when the oxidation process is completed. Consequently, raised mesa structures of single-crystal silicon are formed, surrounded by relatively thick regions of silicon dioxide, as shown in Figure 7.14d. Finally, if the remaining portions of the silicon nitride are removed, it can easily be seen that the surface of the wafer will be extremely flat because the thick oxide regions are actually *recessed* into the silicon substrate with their top surfaces being coplanar with the silicon mesas.

Historically, one of the critical problems previously associated with the fabrication of MOS integrated circuit structures was the fact that relatively thick layers of silicon dioxide were frequently required in the field regions to achieve relatively high values of field-inversion voltage, and, as a result, it became extremely difficult to ensure the continuity and reliability of evaporated metal interconnections that had to climb the relatively sharp steps presented by these thick oxide regions at numerous locations within the device. Wherever the metal was required to cross these steps, localized thinning was observed that could only be avoided by the use of multidirectional or planetary evaporation systems, careful control of step shape, or the use of very thick metal layers. If such precautions were not taken, open-circuits would frequently occur at step locations, and even if continuity were initially observed, the device would still be highly susceptible to long-term failure as a result of

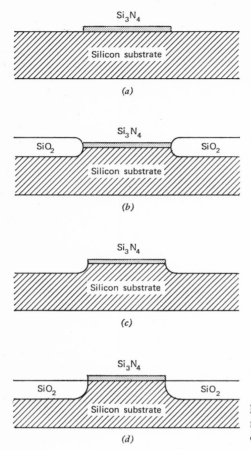

FIGURE 7.14 The use of silicon nitride to selectively oxidize portions of a silicon wafer.

electromigration because of the substantially higher current density in the metal where thinning took place. However, as can easily be seen from the previous discussion, the use of localized oxidation techniques enables one to fabricate MOS integrated circuits with very thick silicon dioxide regions between active silicon devices while virtually eliminating the step-coverage problem once and for all by recessing these regions into the substrate.

A cross-sectional view of an active n-channel silicon-gate MOS transistor along with an n^+ diffused silicon interconnection and a polycrystalline silicon interconnection fabricated through the use of the Coplamos process is shown in Figure 7.15i. (The actual processing steps required to achieve this structure are illustrated in a through h, which are discussed shortly.) All active silicon regions (i.e., drains, sources, channels, and diffused interconnections) are

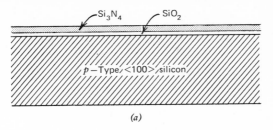

(a)

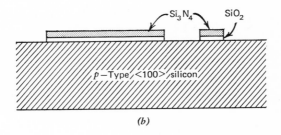

(b)

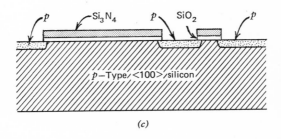

(c)

FIGURE 7.15 Processing sequence for the fabrication of *n*-channel silicon-gate MOS integrated circuit structures through the use of *Coplamos* technology.

located on raised mesas, surrounded by a relatively thick layer of silicon dioxide that overlies the parasitic (nonactive) regions of the structure that are recessed into the substrate. A dielectric sandwich consisting of a thin layer of silicon dioxide and an overlying layer of silicon nitride has been used to mask the active mesa regions from the effects of a low-level *p*-type diffusion or implantation into the parasitic regions and from the subsequent localized thick-field oxidation. In the Coplamos process, not only the formation of the mesas through the use of localized oxidation, but also the selective doping of the silicon surface in the thick-field regions are *both* accomplished with only

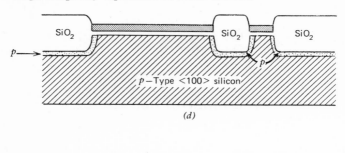

(d)

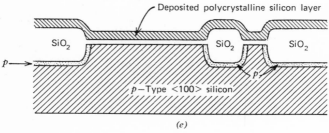

(e)

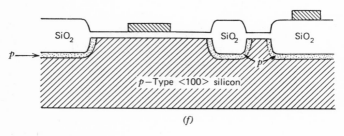

(f)

FIGURE 7.15   (*Contd.*)

one photolithographic operation, ensuring that the *p*-type diffused regions will be self-aligned to and will lie directly under all thick oxide regions within the structure.

The starting material used in the fabrication of the structure shown in Figure 7.15 is typically 3 to 5 Ω-cm (100) orientation *p*-type silicon that is chemically cleaned prior to forming the silicon dioxide-silicon nitride masking layer. The silicon dioxide, which is thin and thermally grown, serves as a stress relief barrier between the silicon and the overlying layer of silicon nitride during the thick field oxidation, and will also subsequently be used as the gate insulator for all active MOS devices. Typically, the thickness of the oxide

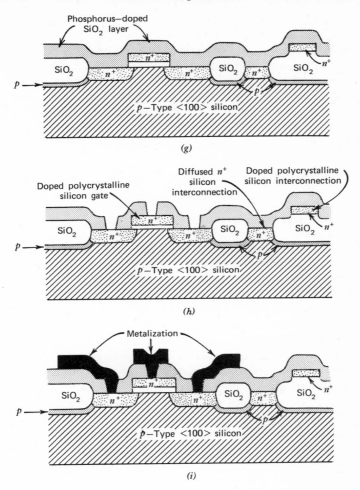

FIGURE 7.15   (*Contd.*)

layer is between 1000 and 1200 Å. The silicon nitride is formed by the decomposition of silicon tetrachloride in ammonia at approximately 850°C, the thickness of the nitride layer being selected to prevent it from cracking at pattern corners as a result of lateral pressure during the prolonged thick field oxidation. A thin layer of silicon dioxide is pyrolytically deposited over the silicon nitride at this point and then densified to facilitate good photo resist adhesion and nitride etch masking.

The first photolithographic operation is now performed, leaving a silicon dioxide-silicon nitride sandwich only in the regions that will eventually become either sources, drains, channels regions, or diffused interconnections. This operation is accomplished by using the patterned resist layer as a mask to etch through the oxide layer with a buffered hydrofluoric acid etch, by removing the resist, by using the etched oxide layer as a mask to selectively etch away the nitride layer in hot phosphoric acid, and finally by etching through the underlying thermally grown oxide layer with the buffered etch using the remaining portions of the nitride layer as a mask. During the latter step, any residual deposited oxide overlying the silicon nitride is also removed. At this step in the process the cross-sectional view of the structure is as shown in Figure 7.15b.

Next, a boron-doped layer of silicon dioxide is pyrolytically deposited over the surface of the wafer at approximately 460°C. The impurity concentration within this oxide must be carefully monitored, since it is to serve as the boron source for the selective outdoping into the thick-field parasitic regions. A short high-temperature drive-in cycle is used to transfer a controlled amount of boron from the doped oxide into the surface of the silicon, after which the doped oxide is etched away (Figure 7.15c).

The localized thermal oxidation of the nonmasked boron-doped field regions that follows is performed in wet oxygen at 975°C. Approximately 10,000 Å of oxide are grown, the upper surface of the oxide being about 5500 Å above, and the lower surface of the oxide being about 4500 Å below the original silicon surface, as shown in Figure 7.15d. The thermal oxidation process converts a few hundred angstroms of the silicon nitride into silicon dioxide and also segregates enough boron into the grown oxide to provide the necessary surface acceptor concentration in the thick field regions. After removing any of the nitride that has been converted into oxide, the remaining regions of silicon nitride are subsequently etched away in hot phosphoric acid, thereby exposing the thin layer of silicon dioxide that will be used for the gate insulator. The wafer is then given a high-temperature nitrogen anneal, and a thin layer of polycrystalline silicon is pyrolytically deposited over the entire surface, as shown in Figure 7.15e. Next, a small amount of the top surface of the polycrystalline silicon layer is thermally oxidized for masking purposes and the second photolithographic operation is performed. After first etching through the masking oxide layer, the underlying poly-crystalline silicon is etched away, as shown in Figure 7.15f, with a mixture of hydrofluoric and nitric acids having an etch rate of approximately 2000 Å/min at room temperature. After this etch, the exposed thin oxide regions are etched away using the remaining polycrystalline silicon areas as a mask.

At this point, a moderately thick layer of heavily phosphorus-doped silicon dioxide is pyrolytically deposited over the entire surface of the wafer, as shown in Figure 7.15g, and the wafer is then placed in a furnace tube at approximately 1050°C for 10 to 15 min in a dry nitrogen ambient. The latter high-temperature heat-treatment accomplishes two things. First, the phosphorus-doped-oxide serves as a diffusion source and enables the n^+ drains, sources, and silicon interconnections to be formed to a junction depth of slightly over 1 μ. At the same time, the silicon gates are also doped. Second, the heat treatment smooths out the contour of the phosphosilicate glass layer by reflowing it over the sharp cliffs presented by the edges of the polycrystalline silicon regions, thereby ensuring the continuity of overlying metal interconnections, as was discussed previously in this section.[30] The wafer is then once again coated with photo resist, and the third photolithographic operation is performed and the contact holes are defined and etched out, as illustrated in Figure 7.15h. After this is accomplished, a thin layer of titanium is evaporated over the surface of the wafer, followed by the evaporation of an overlying layer of thick aluminum. The use of the titanium-aluminum double-layer metalization system was first described by Patterson,[34] and has been shown to be particularly applicable to the fabrication of n-channel MOS integrated circuits because of the ability of the titanium layer to prevent aluminum from "spiking" through dislocations and shorting out shallow n^+ diffused junctions. The wafer is now once again coated with photo resist and the metalization pattern is defined and etched out, as illustrated in Figure 7.15i. The structure is then alloyed in hydrogen at 480°C for approximately 30 min to ensure good metal-to-silicon contacts and to annihilate any fast surface states at the silicon-silicon dioxide interface. Finally, a 5000-Å layer of silicon dioxide is pyrolytically deposited at low temperature over the entire structure, and contact windows are opened photographically over the bonding pad areas.

Through the use of the Coplamos process, it is possible to greatly increase the parasitic field-inversion voltage by selectively doping the silicon surface in the thick-field regions and thereby increase $Q_{SD_{\max}}$ only in these regions. Consequently, while the field-inversion voltage is increased according to (2.44), the effective acceptor doping concentration in the active mesa regions is kept low; thus the channel mobility and the source-body effect associated with the active transistors are not degraded in any way. The capacitance associated with the diffused n^+ regions to substrate is increased because of the higher capacitance per unit area to the p-doped sides of the mesas, but this effect is minimized because of the negative applied substrate bias.

The negative substrate bias, which is typically on the order of a few volts, is used to set the threshold voltages of the active MOSFETs at approximately +1.4 to +2.0 V by means of the source-body effect, as given by (2.80). As was discussed in Section 7.5.1, the use of the negative bias to increase the apparent device threshold voltage is required because of the relatively low acceptor doping concentration present in the active channel regions, which, in this case, results in typical threshold voltages of approximately +0.2 to +0.6 V with zero substrate bias. The source-body effect is linearly proportional to *both* the thickness of the insulator and to the maximum value of the surface depletion region charge density per unit area. The latter quantity is much greater in the thick-field regions than in the active areas because of the selective doping, and the thickness of the silicon dioxide in the thick-field regions is approximately 15,000 Å compared with an active gate insulator thickness on the order of 1000 Å. Thus it follows that the source-body effect associated with parasitic devices is far greater than for active transistors because of the multiplicative effect in (2.80). Consequently, while the selective doping of the thick-field regions results in typical field-inversion voltages of +10 to +20 V, the application of a small negative substrate bias further increases the field-inversion voltage to a level many times greater than the maximum power supply voltage used in the circuit.

The structure shown in Figure 7.15*i* also eliminates space-charge-limited parasitic current flow between nonrelated diffused n^+ regions, making it possible to separate active devices by as little as 0.2 mil and thereby resulting in extremely high device packing density. It can easily be seen here that the selective doping of the silicon in the thick-field regions increases the effective acceptor doping concentration at the surface such that the punch-through voltage, as given by (4.62), is much greater than typical operating power supply voltages even for *drawn* separations of nonrelated diffused regions as small as 0.2 mil. (It should be noted that because of lateral oxidation during the formation of the mesa regions, the separation between mesas will increase by approximately 0.1 mil beyond the original spacing. Also, the length and width of a mesa will shrink by about 0.1 mil for the same reason.)

A further advantage of this technique is that a diode formed between a diffused n^+ region and the *p*-doped field can be used to achieve extremely good protection of thin oxide layers from the effects of static charge rupture. Since the acceptor concentration near the surface in the *p*-doped parasitic regions is sufficiently high to achieve a sharp avalanche breakdown voltage consistently in the +20 to +30-V range, this diode can be used as a protection device without the need for placing a grounded gate electrode over the periphery of a diffused junction, as is commonly done in conventional MOS integrated circuits.

Coplamos technology has recently been used to fabricate *n*-channel silicon-gate high-speed 4096-bit random access memories with access times on the order of 80 nsec and full cycle times on the order of 150 nsec in a chip size approximately 167 mils². Other applications have been in the areas of high-speed MOS data communication circuits and electronic calculator systems.

7.6.6 Fabrication of MOS Field-Effect Transistors and Integrated Circuits with Self-Aligned Gates Through the Use of Ion-Implantation Techniques

Both the molybdenum-gate and the silicon-gate structures rely on the ability of the gate electrode to act as a barrier against the diffusion of impurities into the active channel region of a MOSFET. However, if a combination of both diffusion and ion-implantation techniques is employed to form the source and drain regions, self-aligned MOSFET structures may be fabricated by using the gate electrode as a mask to protect the underlying channel region from incident high-energy impurity ions during the implantation step. Both discrete devices and integrated circuits fabricated in this manner have been found to exhibit extremely small values of parasitic gate-to-drain and gate-to-source capacitances-far less, in fact, than even molybdenum-gate or silicon-gate devices.

The use of ion-implantation techniques to fabricate MOSFETs with self-aligned gate electrodes was first reported by Bower and Dill.[26] Although the entire drain and source regions in these early devices were formed in a single operation by bombarding the wafer with high energy boron ions, they were plagued with large parasitic drain and source resistances because of the high sheet resistivities of the ion-implanted regions and required two separate metalization steps; one step to form the gate regions prior to the implantation operation and another to make contact with the implanted areas after the ion bombardment. However, it soon became apparent that a substantial improvement in device performance could be achieved by combining diffusion and ion-implantation techniques to form the source and drain regions, and employing only one metalization step.[26,35] The resulting fabrication sequence, which was found to be directly applicable to the realization of high-speed, self-aligned p-channel MOS integrated circuits,[36,37] is illustrated in Figure 7.16.

Using conventional thick-oxide aluminum-gate p-channel fabrication techniques, the structure shown in Figure 7.16a is constructed with the gate electrode intentionally positioned *in between* the diffused source and drain regions so that, rather than overlapping these regions, it lies approximately 0.1 to 0.2 mil away from each. Next, the wafer is bombarded with high-energy boron ions. Since both the aluminum gate electrodes and interconnections, and the thick regions of silicon dioxide are equally effective as ion-implantation masks, only the regions of the wafer that are covered by the thin unprotected layer of silicon dioxide will be doped as the ions penetrate through the thin oxide into the surface of the silicon directly below. As shown in Figure 7.16b, the silicon regions between the gate electrode and the diffused p^+ areas are thereby implanted with boron, thus extending the drain

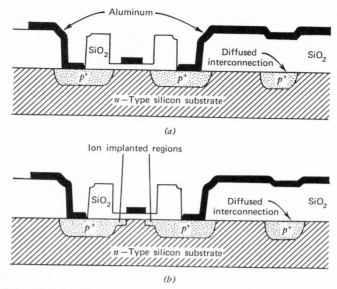

FIGURE 7.16   Fabrication of self-aligned *p*-channel MOSFET through the use of ion-implantation.

and source regions up to the gate electrode and resulting in almost perfect alignment.

Because the lateral movement of the implanted boron ions that penetrate into the silicon substrate is extremely small, the amount of gate-to-drain and gate-to-source overlap for a self-aligned ion-implanted MOSFET will be even less than with self-aligned silicon-gate or molybdenum-gate MOSFETs, where the effects of the lateral diffusion of the drain and source regions on the overlap capacitances will be substantial. This is demonstrated in Figure 7.17, which shows the dependence of the gate-to-drain negative feedback capacitance, or Miller capacitance, on the amount of overlap of the gate electrode over the drain region.[35] The lower abscissa corresponds to a gate oxide thickness of 2000 Å. The bracketed lines represent the range of experimentally measured values for conventional (non-self-aligned) MOSFETs and silicon-gate and ion-implanted devices. It can be readily seen from the experimental data that extremely small values of interelectrode overlap can be achieved through the use of ion-implantation to fabricate MOSFET structures.

After the implantation operation is performed, the radiation damage to the silicon lattice in the implanted regions is annealed out at elevated temperatures in an inert atmosphere. The fraction of implanted boron atoms that become electrically active as acceptors increases with higher annealing temperatures. However, the maximum allowable annealing temperature will

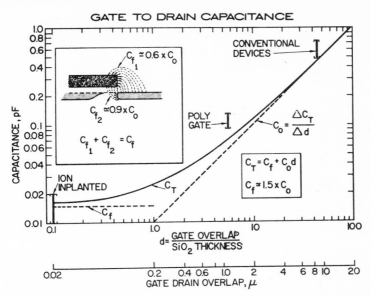

FIGURE 7.17 Miller feedback capacitance produced by the overlap of the drain by the gate. (After Bower, Dill, Aubuchon, and Thompson[35].)

be limited by the melting temperature of the gate electrode. For aluminum-gate devices, the annealing step is performed at approximately 480 to 500°C, and the observed sheet resistivities of the implanted p-regions after the anneal are, typically, about 3000 $\Omega/\square$.

Because of the relatively high sheet resistivity of the implanted p-regions, the distances between the gate electrode and the diffused drain and source regions should be made as small as possible, constrained only by the allowable mask alignment error and the lateral diffusion of the p^+ regions. In this way, the size of such an ion-implanted device will be only slightly greater than that of a conventional MOSFET, and the parasitic drain and source resistances associated with the implanted regions can be kept small. For example, for an ion-implanted MOSFET with a channel width of 10 mils and implanted drain and source regions each 0.1 mil long, the parasitic drain and source resistances will only be 30 Ω, assuming that the sheet resistivity of the implanted boron regions is approximately 3000 $\Omega/\square$. These additional resistances are small compared to typical values of the channel resistance; thus they will not seriously degrade the gain of the device.

On the other hand, the high sheet resistivity of the implanted p-regions enables high-impedance resistors to be incorporated into ion-implanted MOSFET integrated circuits with no additional processing steps.[37] The resistors are implanted in the desired locations at the same time that the drain and source regions of the individual transistors are extended to achieve self-alignment. Contact is made to the resistors through p^+ regions that were previously formed during the diffusion operation. These high-impedance resistors can be used as load elements in inverter stages and can also be used to form precision resistor ladder networks for integrated MOS digital-to-analog converters.

The extremely small values of gate-to-drain and gate-to-source capacitances attainable through the use of ion-implantation have made possible the fabrication of high-speed MOS integrated circuits and high-frequency discrete MOSFETs. Bower, Moyer, and Dill[36,37] have reported the successful fabrication of p-channel ion-implanted integrated MOS shift registers and multiplexers capable of operating at speeds far in excess of 10 MHz. The n-channel ion-implanted MOS devices are particularly attractive for high-frequency applications because of the higher transconductances that result from the high electron mobility. Furthermore, unlike boron, a large fraction of the implanted phosphorus atoms can become electrically activated when the wafer is annealed at 500°C and, consequently, the sheet resistivity of an implanted n-type region typically will be on the order of 300 $\Omega/\square$. As a result, the parasitic drain and source resistances associated with the ion-implanted regions in an n-channel MOSFET can be reduced by approximately an order of magnitude compared to similar p-channel devices.

Using the structure shown in Figure 7.18, Shannon, Beale, Stephen, and Freeman[38,39] were able to fabricate discrete n-channel ultra-high-frequency MOSFETs capable of operating in the gigahertz range by employing two separate ion-implantation operations. A high-resistivity (10 Ω-cm) p-type epitaxial layer of silicon grown to a thickness of approximately 8 μ on a low-resistivity p^+ substrate was used to minimize the parasitic drain-to-substrate capacitance while also reducing the effective drain-to-substrate charging resistance. The devices were constructed with extremely small gate lengths, about 3 μ on the average, to achieve high-transconductance operation. However, because of the small channel lengths and the relatively high resistivity of the p-type epitaxial layer, they were found to be prone to punch-through unless a low-level p-type layer was implanted into the channel region prior to both the metalization step and the subsequent implantation of the n-type regions used to achieve self-alignment. As shown in Figure 7.18, the depth of the p-type implanted layer is shallower than the *diffused* drain and source "contact" regions yet deeper than the junctions formed during the implantation of the phosphorus. The low-resistivity p-type layer at the surface of the silicon resulting from the first implantation acts to retard the spreading of the drain depletion region, thus preventing punch-through and maintaining the high output impedance of the MOSFET under conditions of saturated drain current operation. Also, because of the higher acceptor

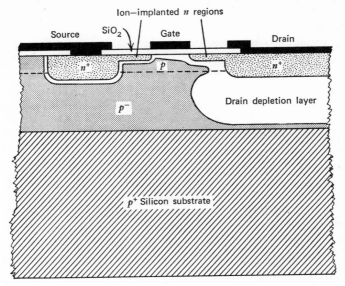

FIGURE 7.18   Cross-sectional view of *n*-channel MOSFET structure for ultrahigh frequency operation fabricated through the use of two ion-implantation operations.

doping concentration in the channel region resulting from the *p*-type implantation, the devices were observed to operate only in the enhancement mode, exhibiting threshold voltages of typically +1.5 V.[39]

As before, any radiation damage that occurred during the implantation steps had to be annealed out in an inert atmosphere at approximately 500°C. Typical devices constructed in the manner above exhibited approximately 4-dB gain at 1 GHz and a maximum frequency of operation of about 1.4 GHz.

7.7   THE USE OF ION-IMPLANTATION TECHNIQUES TO SHIFT THE OBSERVED THRESHOLD VOLTAGES OF MOS FIELD-EFFECT DEVICES

Although the first applications of ion-implantation technology to the fabrication of MOSFETs and integrated circuits were in the areas of reducing the parasitic gate-to-drain and gate-to-source capacitances by using the gate electrode as an implantation mask, and also implanting linear resistors with reproducibly high sheet resistivities, it soon became apparent that perhaps an even more important application was the use of ion-implantation to shift the threshold voltages of MOSFETs. In particular, it was found that the threshold voltage of either an *n*- or a *p*-channel MOSFET could be shifted and controlled over a wide range of values simply by implanting a relatively low concentration of acceptor or donor impurities through the gate insulator of the device and into the underlying silicon substrate.[40,41] Depending on the specific application, the implantation could be performed into the active channel regions of *all* devices within an integrated circuit structure, or only into *selected* devices at the expense of an additional photolithographic operation. The latter technique was shown to be capable of fabricating depletion-mode MOS load elements with enhancement-mode active switching devices on the same monolithic structure, resulting in MOS integrated circuits that were characterized by lower power dissipation, higher switching speeds, greater packing density, and the ability to operate directly off the lower voltage levels normally used to drive bipolar integrated circuits (0 to $+5$ V).[41]

Referring once again to the generalized equation for the threshold voltage of a MOSFET as given by (2.44), it can easily be seen that the threshold voltage can be permanently increased or decreased if, respectively, either a controlled amount of impurities of the *same type* as are found in the silicon substrate, or of the *opposite type*, are implanted into the surface of the silicon in the region separating the drain and source diffusions. Once the implantation is performed, a high-temperature anneal is required to electrically activate the impurities, thereby resulting in either an increase or a decrease in the maximum charge density per unit area contained in the surface depletion region and a corresponding change in the threshold voltage of the device. It should be remembered that the effective impurity doping concentration near the surface in the active channel region after the implantation is performed will no longer be uniform and will vary with depth into the silicon depending on the profile of the implantation and any redistribution of impurities during subsequent high-temperature processing steps.

The ion-implantation operation is usually performed through a thin insulating layer, usually on the order of 1000 Å of silicon dioxide, to prevent surface contamination.[42] The useful range of insulator thickness is limited at the low end by increasing pinhole density and at the high end by run-to-run irreproducibility in the characteristics and profile of the implanted layer.

When monoenergetic ions are implanted into an amorphous material, the concentration profile is a Gaussian curve with the peak located at a distance R_p (the mean projected range) below the surface. The width of the distribution is characterized by the mean standard deviation, ΔR_p. As the energy of the incident ions increases, both R_p and ΔR_p also increase.[40] The irreproducibility that is typically observed in the surface impurity concentration profile when the implantation is performed at low energies through a relatively thick insulator is a direct result of the fact that, under these conditions, the peak of the distribution of the implanted ions lies in the insulating layer

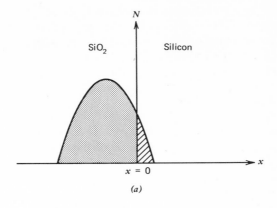

(a)

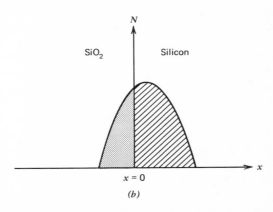

(b)

FIGURE 7.19 (a) Typical distribution of implanted impurity atoms when the implantation is performed at low energies through a relatively thick silicon dioxide layer; (b) typical distribution of implanted impurity atoms with the peak of the distribution located just below the silicon surface when the implantation is performed at higher energies through a thinner silicon dioxide layer.

itself, as shown in Figure 7.19a, rather than in the silicon. Hence only the "tail" of the distribution actually lies in the silicon and even small variations in either the thickness of the insulating layer or in the energy of the incident ions will result in a significant change in the "tail" of the distribution of the implanted impurities. On the other hand, as shown in Figure 7.19b, if the thickness of the insulator is decreased or if the energy of the implantation is increased, the peak of the distribution of the implanted ions can be positioned to lie just below the silicon surface and, under these conditions, small variations in the thickness of the insulator or in the energy of the ions will not have as great an effect on the characteristics of the implanted layer. Dill and Coppen[43] found that if 70-keV boron ions were implanted into a silicon substrate through a layer of silicon dioxide, large variations in the thickness of the oxide would result in only relatively small changes in the observed sheet resistivity of the implanted layer as long as the thickness of the silicon dioxide film was in the range of 500 to 1500 Å. However, for film thicknesses in excess of 1500 Å, they found that the observed sheet resistivity varied greatly as a larger percentage of the implanted ions never reached the silicon and, instead, came to rest in the overlying layer of silicon dioxide.

The amount of impurities actually implanted into the surface of the silicon will be a function of the dose (number of incident ions per unit surface area), the energy of the ions, and the thickness of the silicon dioxide layer. In applications where it is desired to shift the threshold voltage of MOSFETs that have been fabricated with gate oxide layers approximately 1000 Å thick on the order of a few volts, implantation energies of between 15 and 50 keV are usually employed to implant either boron or phosphorus impurities into the active channel region.[43]

Perhaps the most important application of using low-level ion-implantation techniques to shift the threshold voltages of MOSFETs is the fabrication of large-scale-integrated circuits with both enhancement-type active MOSFETs and depletion-type load elements. This can easily be accomplished through the use of conventional fabrication techniques by covering the wafer with photo resist directly after the gate insulating layer has been formed and then exposing and developing the resist such that it will remain only in selected areas to mask the active channel regions of specific devices from the effects of the subsequent implantation. The devices that do not have their channel regions protected by the overlying resist layer will experience a threshold voltage shift because of the implantation. After the implantation operation is performed, the remaining portions of the resist are removed and, once more, conventional processing techniques are employed to complete the fabrication of the devices.

The advantages of achieving both enhancement-type and depletion-type MOSFET characteristics simultaneously on the same monolithic silicon structure can be seen by considering Figures 7.20 and 7.21.[44] Figure 7.20 illustrates a MOS inverter consisting of an enhancement-type active switching device and a depletion-type load element with its gate and source connected in common. If the input to the inverter is such that the enhancement device is initially "on" and the output is approximately at 0 V, when the input is changed to 0 V the enhancement device is then turned "off," and the depletion-mode device must charge the capacitive load at the output to $V_{\text{out}} = V_{DD}$. As can be seen from Figure 7.21, because of its constant-current

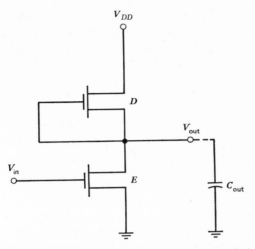

FIGURE 7.20 MOS inverter circuit with depletion-type load element.

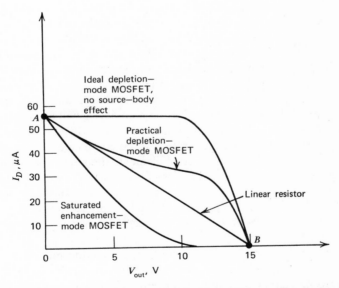

FIGURE 7.21 Comparison of the transfer characteristics associated with depletion-type, linear resistor, and saturated enhancement-type load elements. (After Crawford[44].)

characteristics, an (ideal) depletion-type load element will be able to supply more current as a function of time during the switching cycle as the output swings from point A to point B than with a comparable saturated enhancement load configuration, thus charging the capacitive load faster. This will, of course, result in faster switching speeds. In addition, the use of the depletion-mode load element will permit the output to swing all the way to the supply voltage, V_{DD}.

The potential of the source electrode of the depletion-type device in the inverter configuration of Figure 7.20 will, of course, vary during the switching operation with respect to the fixed substrate potential. Thus it follows that the *source-body* effect will tend to *increase* the pinch-off voltage associated with the load transistor and reduce the magnitude of the saturated drain current flowing through it as the capacitor is charged. This will result in a slight decrease in the switching speed. However, despite this effect, it is evident from the curves of Figure 7.21 that a practical depletion-mode load element will be far superior to a saturated enhancement-mode load and will even outperform a linear resistor load. (Using linear-resistor load elements in large-scale-integrated circuits is impractical from a device fabrication standpoint because of the high impedances required and the extremely large areas that consequently would be needed because of the relatively low sheet resistivities available with most MOS technologies.) If a saturated enhancement-type load element with its gate and drain electrodes connected in common is employed in an inverter configuration, *initially* it can supply the same current to charge the capacitor after the bottom device is switched off. However, as soon as the capacitor begins to charge and V_{out} begins to increase, the gate-to-source voltage for the load device will decrease, causing a substantial reduction in the charging current and thereby increasing the time required for the switching operation to be completed. As was the case for the (nonideal) depletion-mode load element, the source-body effect will also tend to further decrease the charging current as V_{out} increases. Finally, it should be noted that, unlike the case of the depletion-mode load, the output of the inverter cannot charge to the full supply voltage if an enhancement-type device with its gate and drain electrodes tied together is used as the load element.

The many advantages of using ion-implanted depletion-mode load devices with enhancement-mode switching transistors are applicable to both n- and p-channel MOS technologies and have been realized with both conventional and self-aligned structures.[45,46] Such techniques are now being routinely employed to fabricate single chip calculator logic systems, random-access and read-only memories, and a wide variety of custom random logic circuits.

Yet another application of using ion-implantation to control the observed threshold voltages of MOSFETs lies in the fabrication of *complementary*

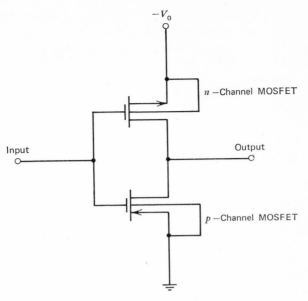

FIGURE 7.22   Complementary MOS inverter circuit.

integrated circuits. Wanlass and Sah[47] first reported that if an enhancement-mode n-channel MOSFET were connected in a series common-gate configuration with an enhancement-mode p-channel MOSFET as shown in Figure 7.22, *the resulting structure could function as a simple inverter and would dissipate virtually no power whatsoever so long as the inverter was in one of its two stable states.* Referring to the complementary inverter circuit illustrated in the figure, the logic signals used are 0 and $-V_0$ V. Since the enhancement-mode p-channel device will have a negative threshold voltage and the enhancement-mode n-channel device will have a positive threshold voltage with respect to their individual source regions, it follows that a signal of 0 V applied to the common input will simultaneously turn the p-channel transistor *off* and the n-channel transistor *on*. In this state, virtually all of the bias voltage will be dropped across the p-channel MOSFET and the output will be approximately $-V_0$ V. On the other hand, an input voltage of $-V_0$ V will turn the n-channel device *off* while turning the p-channel device *on*. The bias voltage will now be dropped across the n-channel transistor, and the output will be approximately 0 V. In either of the two stable conditions described above, one of the MOSFETs will be in a very high-impedance *off* state; consequently, the series combination of the two devices will draw almost no steady-state current. Furthermore, because of the extremely high gate input impedance associated with the MOS structure, no current will flow in the gate circuit under steady-state conditions. Thus it can easily be seen that the complementary MOS inverter will, indeed, dissipate virtually no power when in a stable state. Power will only be dissipated during switching from one state to the other.

The inverter circuit of Figure 7.22 is the basic circuit of complementary MOS logic and is the basic building block of even the most complex complementary MOS large-scale-integrated arrays now being produced today. Besides the extremely low "standby" power consumption, the complementary MOS inverter is also characterized by very high switching speeds and very high noise immunity over a wide range of power supply voltages. The high speed, compared to that of single-channel configurations, is achieved by the push-pull operation of the circuit. During switching from one state to the other, the output capacitance is always charged or discharged through a heavily conducting MOS transistor that has been switched into its low-impedance state. The relatively high noise immunity is a direct result of the symmetry of the inverter configuration, the very high ratio of the off resistance to the on resistance of the individual devices, and the very high gain in the transition region.[48]

It soon became apparent that to fully take advantage of the previously described characteristics of the complementary MOS inverter structure and thereby implement complex high-speed, low-power, large-scale-integrated digital circuits, it was necessary to perfect techniques that would make possible the fabrication of both n- and p-channel enhancement-type MOSFETs on a single monolithic silicon substrate. This, in turn, required the realization of lightly doped p- and n-type regions at specific locations on the surface of a single silicon wafer, the n-channel MOSFETs subsequently being formed in the lightly doped p-type areas and the p-channel MOSFETs being formed in the lightly doped n-type areas. Referring to Chapter 2 and, in particular, to Figures 2.12 and 2.17, it can easily be seen that, for typical values of gate oxide thickness and fixed positive interface charge density, the fabrication of well-matched complementary enhancement MOSFET structures with reproducible threshold voltages consistently in the 1 to 3-V range requires accurately controlling the surface doping concentration in both the lightly doped p- and n-type regions to levels on the order of approximately $10^{16}/cm^3$. Although this was at first extremely difficult, a number of techniques were developed to achieve this end. Yagura, Catlin, and Hutchenson[49] produced sufficiently lightly doped p-regions or "wells" within an n-type silicon wafer through the use of low-level diffusion techniques. The n- and p-channel enhancement-type MOSFETs, with excellent characteristics, were then formed in the opposite conductivity regions, as shown in Figure 7.23a. Similar techniques were subsequently described by Athanas[50] and others. An alternate

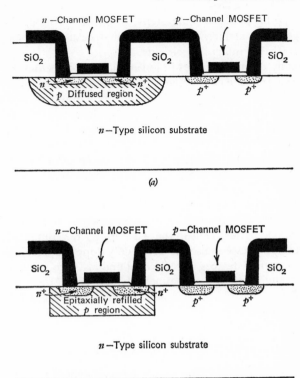

FIGURE 7.23   (*a*) Cross-sectional representation of complementary MOS transistor pair fabricated through the use of the low-level diffusion technique of Yagura *et al.*[49]; (*b*) cross-sectional representation of complementary MOS transistor pair fabricated through the use of the epitaxial refill technique of White and Cricchi[51].

approach to achieving matched complementary MOS transistors on a single monolithic silicon substrate was employed by White and Cricchi.[51] Their technique was based on etching deep moats into the surface of an *n*-type silicon wafer and epitaxially refilling the moats with lightly doped *p*-type silicon. A series of mechanical lapping and polishing operations were then performed to once again achieve a flat wafer surface and the complementary devices were then formed in the resulting opposite conductivity regions within the wafer, as illustrated in Figure 7.23*b*. Although such a procedure is generally less desirable because of the difficulties associated with the mechanical lapping and polishing steps, a similar technique was recently used by Burgess and Daniels[52] to achieve low-threshold-voltage silicon-gate complementary MOS integrated circuits.

With the many recent advances in ion-implantation technology, it became obvious that the opposite conductivity regions required for the fabrication of complementary MOS devices on a single substrate could also be achieved by ion-implantation techniques and the impurity doping concentration at the surface of the implanted regions could be controlled to a far greater accuracy than with low-level diffusion or epitaxial refill techniques, thereby resulting in much better threshold voltage control.[42,43] As an example of the remarkable tolerances that can be achieved with complementary MOS structures through the use of ion-implantation, Coppen, Aubuchon, Bauer, and Moyer[53] described a technique that they employed to fabricate complementary MOS integrated circuits for electronic wristwatch applications for operation off single battery cells in the +1.2 to +1.6-V range. Ion-implantation of boron impurities was used to accurately dope the p-regions or "wells" in which the n-channel devices were formed and again to lower the threshold voltages associated with the p-channel MOSFETs to their desired value. In this way, they were able to consistently achieve threshold voltage control to within 0.6 ± 0.3 V.

7.8 SILICON-ON-SAPPHIRE MOS FIELD-EFFECT DEVICES

Although a great deal of the effort that has been devoted to the development of higher-speed MOSFETs and integrated circuits has been concentrated in attempts to fabricate structures with very small values of parasitic gate-to-drain and gate-to-source capacitances, it must be remembered that a considerable improvement in high-speed performance can also be realized through the reduction of both parasitic junction capacitances and parasitic interconnection capacitances to substrate. This can be accomplished with silicon-on-sapphire MOS devices.

In 1964, Manasevit and Simpson[54] demonstrated that thin single-crystal films of p-type silicon could be grown over highly polished insulating sapphire substrates. Shortly thereafter, Mueller and Robinson[55] reported the use of such films to fabricate n-channel MOSFETs and, by ensuring that the n^+ drain and source regions were diffused all the way down to the silicon-sapphire interface, Ross and Mueller[56] succeeded in virtually eliminating all drain-to-substrate and source-to-substrate capacitances in these structures. Because the n^+ regions extended down to the insulating substrate, only the "side-wall" area of the drain and source diffusions contributed to the parasitic junction capacitance. Since the thickness of the silicon film in a typical silicon-on-sapphire structure is usually about 1 μ, it can easily be seen that a substantial decrease in the parasitic junction capacitances associated with all diffused regions results.

In integrated circuit applications, silicon-on-sapphire technology makes it possible to achieve higher speed operation through the virtual elimination of all parasitic junction and interconnection capacitances to substrate. It should be noted that this is possible only because of the relatively high carrier mobilities obtainable in good-quality silicon-on-sapphire films. Recently, hole and electron surface mobilities comparable to those that have been achieved in MOS structures on bulk silicon substrates have been reported for silicon-on-sapphire devices.

By selectively etching away all unwanted areas so that the silicon film remains only in the regions where active MOS transistors and diffused interconnections are to be formed, aluminum interconnections between different devices can be routed directly over and in direct contact with the sapphire substrate if desired. The sapphire is an extremely good insulator; consequently, the capacitance to substrate associated with these interconnections is virtually zero. Also, the possibility of spurious metal-to-substrate short-circuits is eliminated. Furthermore, unlike conventional MOS integrated circuits fabricated in bulk silicon, because of the removal of the silicon from all nonactive regions in the silicon-on-sapphire structure, no parasitic transistor action can take place between neighboring active MOSFETs.

7.8.1 p^+p-p^+ Deep-Depletion Silicon-on-Sapphire MOSFETs

Because of the thin-film nature of the silicon-on-sapphire MOSFET structure, Hofstein[57] and Heiman[58] found it possible to fabricate "junctionless" enhancement devices by diffusing p^+ drain and source regions into thin high-resistivity p-type silicon-on-sapphire films. As shown in Figure 7.24, if the resistivity of the silicon film is sufficiently high and if the film is sufficiently thin, the surface depletion region with the gate-to-source voltage set equal to zero can extend all the way through the silicon to the sapphire substrate. Consequently, conduction between the drain and the source will be cut off in a manner quite similar to the operation of a junction field-effect transistor. However, unlike the situation in a junction field-effect transistor, because of the presence of the insulated-gate structure, the device may be operated in the enhancement mode. With the application of a negative gate voltage, a p-type accumulator layer will form in the channel region and appreciable drain-to-source conduction will be observed. The device will be of the *enhancement* type only if the width of the surface depletion region with zero gate voltage is greater than the thickness of the silicon film:

$$x_d \,(V_G = 0) > x_{\text{silicon}}. \tag{7.34}$$

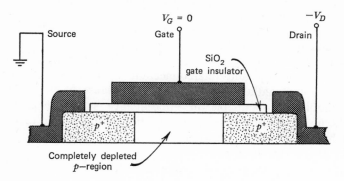

FIGURE 7.24 Cross-sectional representation of a p-channel deep-depletion silicon-on-sapphire MOSFET.

This is achievable with p-type silicon films of approximately 10 Ω-cm or greater grown to thicknesses of less than a micron.[59]

The device will be of the *depletion* type if

$$x_d (V_G = 0) < x_{\text{silicon}}, \tag{7.35}$$

and will exhibit complete pinch-off of drain-to-source conduction with the application of a positive gate-to-source voltage only if the maximum width of the surface depletion region is greater than the thickness of the silicon film:

$$x_{d_{\max}} > x_{\text{silicon}}. \tag{7.36}$$

Thus if an inversion layer forms at the silicon surface before the channel is completely depleted, then total pinch-off of the drain-to-source current will not be observed. However, channel pinch-off at the drain will always occur; hence drain current saturation will always be observed.[57] Because of the high resistivity of the p-type silicon films, the drain-to-source spacing in these devices must be kept sufficiently large to avoid unwanted space-charge limited current flow between drain and source under normal operating conditions.

Deep-depletion n-channel MOSFETs can easily be fabricated by diffusing n^+ drain and source regions into thin high-resistivity n-type silicon films that have been formed on an insulating sapphire substrate.[58] While the considerations discussed above relate to the operation of p-channel deep-depletion devices, the characteristics of the n-channel devices are quite similar and their operation can be described by analogy. The n-channel devices will always be of the depletion type so long as aluminum is used for the gate electrode, since the charge density in the surface depletion region will be a *positive* quantity. Total pinch-off of the drain current with increasingly negative gate voltage will occur only if the silicon film is sufficiently thin so that (7.36) can be satisfied.

7.8.2 Complementary Silicon-on-Sapphire MOS Field-Effect Transistors

It has been shown that both *n*- and *p*-channel (deep-depletion) MOSFETs can be fabricated in the same *p*-type silicon-on-sapphire film. It follows that if the resistivity and thickness of the silicon film and the magnitude of the fixed positive charge density per unit area at the oxide-silicon interface, Q_{SS}, are chosen so that (7.34) is satisfied and neither a *p*-type accumulation layer nor an *n*-type inversion layer is formed at the silicon surface with zero gate voltage, then complementary *n*- and *p*-channel MOS enhancement devices may be fabricated simultaneously on the same wafer using silicon-on-sapphire technology. This was first successfully demonstrated by Allison, Burns, and Heiman.[59,60] Unlike the more conventional approaches to the fabrication of complementary MOS integrated circuits that were previously discussed in Section 7.7, the use of this technique eliminates the need for an extra diffusion or implantation step to form the *p*-type low-conductivity wells in which the *n*-channel devices are to be placed. Besides the many advantages of silicon-on-sapphire structures, integrated circuits fabricated in this manner also benefit from the faster switching speeds and low standby power dissipation usually associated with complementary MOS systems. The processing steps that are normally employed to fabricate complementary silicon-on-sapphire devices are illustrated in Figure 7.25. First, a high-resistivity *p*-type silicon film, usually on the order of 10 to 15 Ω-cm, is grown on a sapphire substrate. The single-crystal silicon film is then covered with a layer of silicon dioxide that is deposited over the wafer at approximately 400°C. The silicon dioxide is patterned and etched and the remaining areas of silicon dioxide are used as a mask to define the active regions of the circuit while the undesired regions of silicon are etched away. After this is accomplished, the remaining silicon dioxide is removed and a layer of boron-doped silicon dioxide is deposited at low temperature over the surface of the wafer. The boron-doped oxide is then patterned and etched away except in the regions where the *p*-channel devices will be formed. Next, a layer of phosphorus-doped silicon dioxide is deposited over the wafer, again at a low temperature, and it is similarly patterned and etched away except in the regions where the *n*-channel devices will be formed. The channel regions for both the *n*- and *p*-channel devices are defined by simultaneously *selectively* etching away the overlying phosphorus-doped and boron-doped oxides, respectively. Now the wafer is subjected to its only high-temperature processing step as the insulating gate

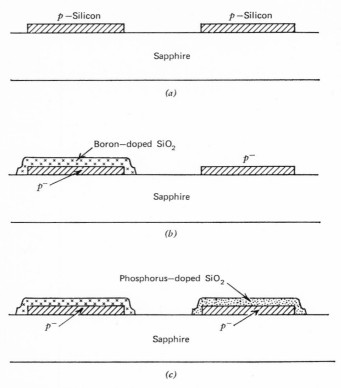

FIGURE 7.25 Fabrication sequence for the construction of silicon-on-sapphire MOS complementary devices.

oxide regions are thermally grown over the channel areas. At the same time, the drain and source regions for both the n- and the p-channel devices are formed by outdiffusion from the doped-oxide layers. Contact windows to the diffused regions are then defined and etched, aluminum is deposited over the wafer, and the metalization pattern is defined and etched. The final structure shown in Figure 7.25f, as previously noted, is characterized by extremely small junction and metal interconnect capacitances to substrate, and by reduced gate-to-drain and gate-to-source interelectrode capacitances because of the doped-oxide outdiffusion technique.[61] Complementary MOSFET integrated circuits fabricated in the manner above have been shown to be capable of operating at speeds many times greater than conventional single-channel MOS devices fabricated on bulk silicon substrates. For example, Meyer, Burns, and Scott[62] have reported complementary MOS silicon-on-sapphire 50-bit shift registers that were able to operate at speeds above 75 MHz.

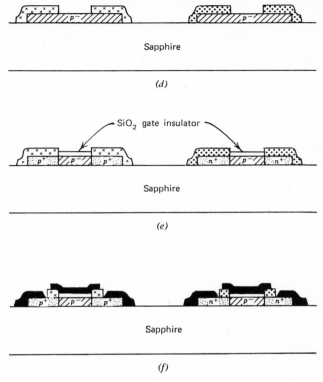

Sapphire

(d)

SiO$_2$ gate insulator

Sapphire

(e)

Sapphire

(f)

FIGURE 7.25 *(Contd.)*

It should be pointed out that conventional (bulk) silicon device fabrication techniques are *not* directly applicable to the construction of high-quality silicon-on-sapphire MOS integrated circuits. In particular, because of the possibility of contamination of the silicon film by outdiffusion of aluminum and other elements from the sapphire substrate, and because of the high degree of disorder in the silicon near the sapphire-silicon interface that results in accelerated diffusion, accelerated silicon etch rate, and accelerated oxidation effects near this interface, it has been shown[61] that the minimization of the time during which a wafer is exposed to high-temperature processing steps and the elimination of thermal oxidation steps whenever possible are necessary precautions to ensure reproducible, high-quality silicon-on-sapphire MOS devices. This is especially true when the structures are fabricated using the previously discussed deep-depletion technology. Recently, however, another approach to the fabrication of complementary MOS silicon-on-sapphire devices has been reported, which although it involves a more complex processing sequence, is less susceptible to some of the problem areas encountered when employing p-type deep-depletion MOSFETs to form complementary devices according to the fabrication sequence illustrated in Figure 7.25.[63,64] Rather than employing only one (high-resistivity p-type) silicon film to fabricate both types of devices, this alternate approach relies on a two-stage epitaxial deposition process that results in two low-resistivity (p- and n-type) silicon films in which the n- and p-channel devices, respectively, are separately fabricated. Since the doping concentrations in both the p- and n-type films are significantly greater than the doping concentrations required for the deep-depletion technology, the complementary MOS devices fabricated with the two-stage epitaxial deposition technology exhibit higher threshold voltages, but are characterized by a much greater tolerance to doping variations in the silicon films. Circuits operating at speeds up to 95 MHz have already been designed and constructed through the use of this technology.[63] Finally, it should be noted that although it is quite desirable to be able to achieve both n- and p-channel enhancement-type devices on the same structure through the use of silicon-on-sapphire technology, virtually all of the advantages of such a configuration will be retained if the n-channel device has a slightly negative threshold voltage and is therefore of the depletion-type. The only disadvantage will be an increase in the standby power dissipation of each inverter within the circuit resulting from the fact that a small residual amount of current will still flow through the n-channel MOSFET even when its gate-to-source voltage is set equal to zero. Rapp and Ross[65] have reported that structures consisting of p-channel enhancement-mode MOSFETs with n-channel depletion-mode (deep-depletion) MOSFETs can easily be fabricated by using a thin *n-type* silicon-on-sapphire film and following a processing sequence similar to the one illustrated in Figure 7.25.

7.8.3 Silicon MOS Field-Effect Devices Fabricated on Other Insulating Substrates

Although a great deal of work has been done to date using silicon-on-sapphire MOSFETs, it must be stressed that the many advantages of this technology are not unique only to silicon-on-sapphire structures, but are also obtainable through the use of other insulating substrates. While extremely good results have already been reported with silicon-on-sapphire structures, with devices now becoming commercially available, and while improvements and refinements are continually being made, some promising results have also been achieved with silicon thin-film MOSFETs that have been formed on other insulating substrates.

The semiconducting properties of silicon thin films that have been epitaxially grown on single-crystal sapphire (α-Al_2O_3) substrates seem to be inherently limited by the crystallographic mismatch between the rhombo-hedral sapphire lattice and the cubic silicon lattice, and by the chemical activity of the sapphire itself.[66] The lattice mismatch at the silicon-sapphire interface leaves the deposited silicon with a distorted crystal structure near this boundary, although the crystal structure of the silicon becomes more nearly ideal as the distance from the sapphire increases. As a result of the variation in crystal structure throughout the deposited silicon film, accelerated diffusion, accelerated oxidation, and accelerated silicon etch rate have been observed near the silicon-sapphire interface,[61] thus complicating the proce-dures that must be employed to fabricate silicon-on-sapphire devices. In addition, aluminum outdoping from the sapphire substrate can seriously affect both the resistivity and the carrier mobilities associated with the silicon film.

Magnesium aluminate spinel ($MgO:xAl_2O_3$) has been found to be espe-cially well suited as a substrate for the epitaxial growth of single-crystal silicon thin films.[67] Spinel has a cubic lattice structure that is very similar to that of silicon and exhibits much less contamination of the silicon film as a result of outdoping during high-temperature processing steps in comparison to sapphire.[67,68] Spinel has a rather wide chemical composition range (x can vary approximately between 0.64 to 6.7) yet only the low-aluminum-rich spinels ($1.5 \leq x \leq 2.5$) are both thermally stable under typical high-tempera-ture processing conditions and also soft enough to permit relatively simple surface preparation techniques to be used prior to device fabrication.[69] The successful construction of n-channel silicon-on-spinel MOSFETs employing silicon dioxide gate insulators has already been reported.[69] These devices exhibited electron field-effect mobilities as high as 420 cm²/V-sec. Metal-insulator-semiconductor field-effect transistors employing both Si_3N_4 and Al_2O_3 gate insulators have also been achieved on spinel sub-strates.[70]

Besides sapphire and spinel, researchers have also been able to grow single crystal silicon thin films on both beryllium oxide and polycrystalline silicon substrates. Beryllium oxide is of particular interest because of its high thermal conductivity that could enable small, highly complex, future high-frequency integrated circuits to dissipate large amounts of power in a relatively small amount of chip area. However, at the present time, the growth of single crystal beryllium oxide substrate material is very difficult and relatively little material is available for experimentation.

7.9 MOS FIELD-EFFECT DEVICES WITH VERY SMALL CHANNEL LENGTHS

Since the maximum frequency of operation of an insulated-gate field-effect transistor is directly proportional to the ratio of its transconductance to the active channel capacitance from gate to channel when the effects of any parasitic capacitances associated with the device structure have been made small, it follows that the maximum frequency at which the device will operate may be increased by decreasing the channel length of the device, since this will both increase its transconductance and decrease the active channel capacitance. A number of complications can arise, however, when extremely small channel lengths are desired. For instance, if the doping concentration in the silicon substrate is not sufficiently high, punch-through can occur between the drain and source for small values of applied drain voltage. Besides seriously degrading the drain saturation resistance of the transistor, this will be particularly undesirable if the device is to be used in a digital application, because it may prevent an "off" state from being observed at typical values of applied drain voltage when the gate voltage is set equal to zero. Furthermore, carrier velocity saturation effects will be observed in devices with very small channel lengths because of the high electric field strengths that will result between drain and source under typical operating conditions. The onset of carrier velocity saturation imposes constraints on the maximum transconductance that can be achieved by a given device structure, which is treated in the next section.

7.9.1 Electrical Characteristics of MOS Field-Effect Devices Operating under Saturated Carrier Velocity Conditions

It is well known that, for very high electric fields, the carrier mobilities observed in silicon and other semiconductors *decrease* with increasing field strength.[71–73] In general, this will occur when the electric field is so high that, during the time between collisions, the carriers are accelerated to speeds many times greater than their thermal speeds. Under these conditions, they are commonly referred to as "hot carriers" because of their relatively high kinetic energies. Beyond a critical value of applied electric field strength, a number of researchers have reported an $\mathscr{E}^{-1/2}$ dependence of carrier mobilities in silicon and germanium and, for extremely high field strengths, have observed an $\mathscr{E}^{-1}$ dependence where complete carrier velocity saturation occurred. Hofstein[74] has pointed out that if all the carriers in the channel region of a MOSFET exhibited complete velocity saturation, an expression for the charge in the channel could no longer be obtained from a one-dimensional solution of Poisson's equation and, because of the need to solve Poisson's equation in two dimensions, no general analytic solution for the channel charge distribution would be possible. Instead, Hofstein treated a simplified model that assumed that velocity saturation would first occur only at the drain end of the channel (where the transverse electric field is of highest intensity) while the remaining portion of the channel could still be characterized by a one-dimensional solution of Poisson's equation. Following Hofstein's treatment, the channel of a MOSFET can be divided into two regions. Region 1, which extends from the source at $y = 0$ to some point $y = y_1$, is a region where the carriers have not reached the saturated drift velocity. It is assumed that, because of the high field strength near the drain, the carriers in region 2, which extends from $y = y_1$ to $y = L$, have reached the saturated drift velocity. If the saturation value of the drift velocity of the carriers is denoted by v_1, then the drain current per unit channel width will be given by

$$I_{DW} = v_1 \sigma \; (y = L), \tag{7.37}$$

where $\sigma \; (y = L)$ is the mobile charge density per unit area at the drain end of the channel. If the drain-to-source separation L is sufficiently small, the region of saturated carrier velocity will form adjacent to the drain region at a drain voltage well below pinch-off. The formation of this region affects the drain current in much the same way as when the drain depletion region pinches off the channel of a MOSFET with a long channel length. That is, current saturation is observed for increasing applied drain voltage. However, if the channel length of the device is such that velocity saturation occurs near the drain at approximately the same voltage required to achieve pinch-off near the drain end of the channel, it is difficult to assess which mechanism will be responsible for the current saturation effect. When L is sufficiently small that a region of saturated carrier velocity forms near the drain at a value of drain voltage significantly less than the pinch-off voltage, the saturated current characteristics of the device will be quite different from the square-law dependence usually observed beyond pinch-off in MOSFETs with longer channel lengths. In particular, under these conditions, it will be shown that for gate voltages substantially greater than the threshold voltage, the transconductance in the region of saturated drain current will be approximately independent of both the applied gate voltage and the drain-to-source separation L.

It will be assumed that, for a given gate voltage, the drain voltage is slowly increased until a region of saturated carrier velocity just starts to form at the drain end of the channel (i.e., $y \approx L$). The drain voltage at which this occurs will be denoted by V_{D1} and the critical field at $y = y_1$ when velocity saturation occurs will be denoted by $\mathscr{E}_1$. If V_{D1} is much less than $(V_G - V_T)$, the drain current per unit channel width will be approximately given by

$$I_{DW} \cong \frac{\epsilon_{ox}\mu}{T_{ox}y_1}\left[(V_G - V_T)V_{D1} - \frac{V_{D1}^{\,2}}{2}\right]. \tag{7.38}$$

The critical field at $y = y_1$ can be expressed as

$$\mathscr{E}_1 = \frac{v_1}{\mu} = \frac{I_{DW}}{\mu\sigma\,(y = y_1)} \cong \frac{\epsilon_{ox}[(V_G - V_T)V_{D1} - \frac{1}{2}V_{D1}^{\,2}]}{T_{ox}L\sigma\,(y = y_1)}, \tag{7.39}$$

and the mobile charge density at $y = y_1$ which is required to terminate the electric field across the gate insulator is given by

$$\sigma_{(y=y_1)} = \epsilon_{ox}\mathscr{E}_{ox} = \frac{\epsilon_{ox}(V_G - V_T - V_{D1})}{T_{ox}}. \tag{7.40}$$

Substituting (7.40) into (7.39) yields

$$\mathscr{E}_1 \cong \frac{[(V_G - V_T)V_{D1} - \frac{1}{2}V_{D1}^{\,2}]}{y_1(V_G - V_T - V_{D1})} \tag{7.41}$$

for $(V_G - V_T) \ggg V_{D1}$. Solving (7.41) for the drain voltage at which carrier velocity saturation occurs at the drain end of the channel gives

$$V_{D1} \cong (V_G - V_T) + \mathscr{E}_1 y_1 - [(V_G - V_T)^2 + \mathscr{E}_1^{\,2} y_1^{\,2}]^{1/2}. \tag{7.42}$$

The drain current per unit channel width may now be obtained by substituting (7.42) into (7.38). The result is

$$I_{DW} \cong \frac{\epsilon_{ox}\mu}{T_{ox}y_1}(\mathscr{E}_1^{\,2} y_1^{\,2})\left\{\left[1 + \left(\frac{(V_G - V_T)^2}{\mathscr{E}_1^{\,2} y_1^{\,2}}\right)\right]^{1/2} - 1\right\}, \tag{7.43}$$

and since y_1 is approximately equal to the channel length L for the specific case where the region of saturated carrier velocity has just started to form at the drain end of the channel, the final expression for the drain current per unit channel width is approximately given by

$$I_{DW} \cong \frac{\epsilon_{ox}\mu}{T_{ox}L}(\mathscr{E}_1^{\,2} L^2)\left\{\left[1 + \left(\frac{(V_G - V_T)^2}{\mathscr{E}_1^{\,2} L^2}\right)\right]^{1/2} - 1\right\}. \tag{7.44}$$

Assuming that the drain current saturates when carrier velocity saturation occurs near the drain end of the channel, it follows that (7.44) is valid for all drain voltages greater than V_{D1} and below the drain breakdown voltage. The transconductance of the device in the region of saturated current flow under these conditions can be obtained by differentiating (7.44) and multiplying by the channel width W:

$$g_m \cong \frac{\epsilon_{ox}\mu W(V_G - V_T)}{T_{ox}L\{1 + [(V_G - V_T)/\mathscr{E}_1 L]^2\}^{1/2}}. \tag{7.45}$$

It is interesting to note the limiting behavior of (7.45) with increasing gate voltage. For gate voltages sufficiently above the threshold voltage such that

$$\left(\frac{V_G - V_T}{\mathscr{E}_1 L}\right)^2 \gg 1, \tag{7.46}$$

the transconductance approaches

$$g_m \rightarrow \frac{\epsilon_{ox} \mu W \mathscr{E}_1}{T_{ox}}. \tag{7.47}$$

Thus on the basis of the model of carrier velocity saturation occurring at the drain end of the channel and, in turn, causing saturation of the drain current at values of drain voltage well below the expected pinch-off voltage, *one would expect the transconductance in the saturated current region to be independent of both the channel length of the device and the applied gate voltage,* for gate voltages that satisfy (7.46). This behavior, in fact, has been observed experimentally in MOSFETs which have been fabricated with very small channel lengths, as discussed in the following sections. Finally, it should be noted that since (7.47) predicts that the transconductance of a MOSFET that is operating under velocity saturated conditions will be independent of the gate voltage, it follows that in such a device there is no need to drive the MOSFET at higher gate voltages to increase its transconductance and its maximum frequency of operation. Instead, maximum transconductance can be achieved as soon as the gate voltage is sufficiently high to satisfy (7.46) and high-speed operation can thereby be obtained at lower power levels.

7.9.2 Carrier Velocity Saturation Effects in Self-Aligned Ion Implanted MOSFETs with Micron Channel Lengths

As discussed in Section 7.6.6, Shannon and Beale were able to fabricate high-frequency self-aligned n-channel insulated-gate MOSFETs with extremely small channel lengths through the use of ion-implantation techniques.[38] These devices, which were constructed with channel lengths of approximately 3 μ, exhibited carrier velocity saturation effects that limited their maximum frequency of operation. More recently, Fang and Crowder[75] reported the successful fabrication and operation of n-channel enhancement ion-implanted MOSFETs with 1-μ channel lengths that exhibited electrical characteristics consistent with the predictions of the carrier velocity saturation model treated in the previous section. The devices were fabricated by diffusing n^+ drain and source regions into a high-resistivity p-type (100) silicon substrate. Then, the channel region of each device was subjected to a low-level boron implantation (followed by a high-temperature anneal) to retard the spread of the drain depletion region at the silicon surface with increasing drain voltage so that punch-through could not be experienced under normal operating conditions. The depth of the boron implant was less than the junction depth associated with the diffused n^+ drain and source regions to reduce the drain-to-substrate junction capacitance. Next, the silicon dioxide gate insulator was grown and a thin, 1-μ wide aluminum gate electrode was positioned between the diffused regions and, at the same time, the aluminum contacts to the source and drain regions were formed. The thickness of deposited aluminum layer was limited to about 5000 Å so that the tight dimensional control required to etch the one micron gate length could be achieved. Finally, the wafer was subjected to a heavy implantation of phosphorus ions, thereby completing the conducting links between the active channel region and the diffused n^+ drain and source regions. A subsequent anneal for 20 min at 514°C lowered the sheet resistivity of the phosphorus-implanted areas to approximately 60 $\Omega/\square$. Measurements from 3.0 to 7.5 GHz indicated that the devices were capable of delivering gain up to a maximum frequency of operation of between 10 to 14 GHz. Approximately 9 dB of gain was observed at slightly less than 4 GHz. It is of particular interest to note that carrier velocity saturation was observed in these devices under typical operating conditions and, as predicted in the previous section, the transconductane in the region of saturated drain current was found to be approximately independent of both the gate voltage and any variations in the actual channel length. Electron velocity saturation in these devices occurred at a critical transverse electric field strength of approximately 5×10^4 V/cm. The actual maximum velocity of the electrons in the channel was found to be on the order of 6.5×10^6 cm/sec at room temperature.

7.9.3 Diffusion-Self-Aligned Techniques for the Fabrication of MOSFETs with Micron Channel Lengths

The concept of using a diffusion-self-aligned structure to fabricate n-channel enhancement-type MOSFETs with effective channel lengths on the order of 1 to 2 μ was first proposed and successfully demonstrated by Harris.[76] Harris constructed the devices on high-resistivity p-type silicon substrates by first selectively etching through a masking layer of silicon dioxide to define only the window for the source diffusion, and then performing a low-level p-type (boron) diffusion followed by a prolonged drive-in that served to both lower the effective doping concentration at the silicon surface and also to drive the p-region laterally beyond the edges of the source window. Next, the drain region was defined by again etching through the silicon dioxide masking layer with the source window remaining open. At this point, a high-concentration n^+ phosphorus diffusion was performed to form the drain and source regions, and the device was completed in a conventional fashion using standard MOS fabrication techniques. Since the low-level p-diffusion was driven laterally beyond the source region, and since the remainder of the surface of the silicon between the two n^+ regions was heavily inverted even at zero gate voltage because of the relatively high resistivity of the p-type substrate, the *effective channel length* of the device became purely a function of how much further the p-type diffused region had spread laterally with respect to the edge of the diffused n^+ source region. Typically, the devices exhibited channel lengths on the order of 1.5 μ. Although the diffusion-self-alignment technique, in contrast to the ion-implantation methods described in previous sections, requires no extremely critical etching operations to define micron linewidths, since the channel length itself is purely determined by the diffusion profiles of the p^- and n^+ diffusions, and since the channel is automatically self-aligned to the source region by the double diffusion out of the source window, extremely accurate control must be maintained on the final surface concentration of the p-diffusion that will be particularly important in determining the threshold voltage of the device.

As shown in Figure 7.26, the same processing technique can be easily extended to fabricate n-channel MOS integrated circuits with enhancement-type active devices and depletion-type load elements.[77] The advantages of such a circuit are well-known and have been treated by Lin and Varker.[78] Tarui, Hayashi, and Sekigawa[77] used this approach to fabricate high-speed MOS NOR gates with depletion-type load elements that were characterized by switching times on the order of 1 nsec when operated with a +5 V power supply and a capacitive load corresponding to a fan-out of 3. Typical threshold voltages for the active MOSFETs were on the order of +2 V. Because of the submicron channel lengths associated with the individual active MOSFETs, and since the devices were fabricated on a high-resistivity p-type silicon substrate, very high transconductances and low "on" resistances were observed, along with low junction capacitances to substrate. Since a p-type substrate was used, the channel region was not floating but was in direct ohmic contact to the substrate. The fabrication technique that was employed, as shown in Figure 7.26, required one additional photolithographic operation and an additional diffusion step beyond conventional MOS integrated circuit technology.

A similar device, which also employed the diffusion-self-alignment technique but was fabricated on an n-type silicon substrate, was reported on by Cauge and Kocsis.[79] A cross-sectional view of this device is shown in Figure 7.27. While the techniques used to construct the structure were very similar to those employed by Tarui, Hayashi, and Sekigawa, the use of the n-type substrate allowed the depletion region associated with the drain-channel junction to spread into the higher-resistivity n-region rather than into the channel region, thereby removing the maximum drain voltage limitation that

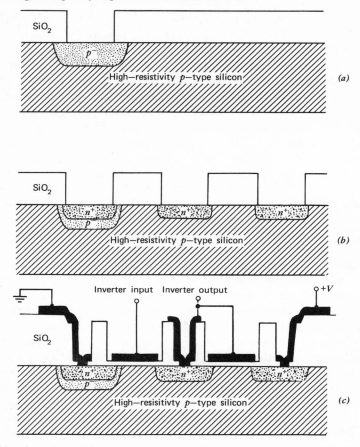

FIGURE 7.26 Fabrication of *n*-channel enhancement-type MOSFET with depletion-type load through the use of diffusion-self-alignment.

would normally be imposed on the device by punch-through considerations, and also resulting in a very small gate-to-drain feedback capacitance. The devices were observed to be operating in a carrier-velocity-saturated mode and the electrical characteristics of the devices in the region of saturated drain current were found to be very similar to those observed by Fang and Crowder.[75] It can easily be seen that the MOSFET structure shown in Figure 7.27 would require some form of isolation between individual devices for integrated circuit applications, since, if this were not done, the drain regions of all the transistors would be in direct ohmic contact with each other. Cauge and Kocsis also described the fabrication of these devices using both thin-film

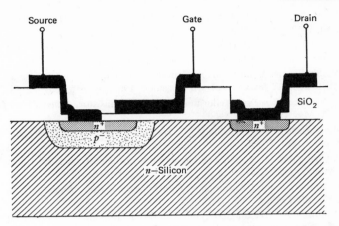

FIGURE 7.27 Cross-sectional view of a diffusion-self-aligned *n*-channel MOSFET for high-frequency applications that has been fabricated on an *n*-type silicon substrate.

silicon-on-insulator structures and p^- substrates with thin epitaxially grown n^- silicon films to provide isolation between the individual devices. A side benefit of the latter technique was that if the low-level *p*-diffusion out of the source window was allowed to extend through the *n*-epitaxial layer and into the *p*-type substrate, the channel region was then able to make ohmic contact to the substrate, and, as a result, the channel would no longer be floating. Junction isolation could be provided by positioning other p^- diffused regions around the individual transistors in a manner quite similar to the technique commonly used to isolate common-collector regions in bipolar epitaxial integrated circuits. Cauge and Kocsis reported that the devices exhibited considerable gain at frequencies well above 1 GHz. The switching speeds of the transistors were measured for inverter configurations. Rise times of 0.48 nsec and delay times of 0.27 nsec were observed.

REFERENCES

1. P. Richman, *Characteristics and Operation of MOS Field-Effect Devices*, McGraw-Hill Book Co., New York, 1967, pp. 101–108.

2. H. Yamamoto, M. Shiraishi, and T. Kurosawa, A Forty Nanosecond, 144-Bit *n*-Channel MOS IC Memory, presented at the 1969 International Solid State Circuits Conference, Philadelphia, Pennsylvania, February 1969.

3. P. Richman and J. Hayes, Coplamos Keeps *n*-Channels in Line, *Electronics*, Vol. 45, No. 10, May 1972, pp. 111–114.

4. P. K. Weimer, The TFT-A New Thin-Film Transistor, *Proceedings of the IRE*, Vol. 50, 1962, pp. 1462–1469.

5. S. R. Hofstein and F. P. Heiman, The Silicon Insulated-Gate Field-Effect Transistor, *Proceedings of the IEEE*, Vol. 51, September 1963, pp. 1190–1202.

6. F. V. Shallcross, Evaluation of Cadmium Selenide Films for Use in Thin-Film Transistors, *RCA Review*, Vol. 24, December 1963, pp. 676–687.

7. P. K. Weimer, A *p*-Type Tellurium Thin-Film Transistor, *Proceedings of the IEEE*, Vol. 52, May 1964, pp. 608–609.

8. H. A. Klasens and H. Koelmans, A Tin Oxide Field-Effect Transistor, *Solid State Electronics*, Vol. 7, 1964, pp. 701–702.

9. L. L. Chang and H. N. Yu, The Germanium Insulated-Gate Field-Effect Transistor, *Proceedings of the IEEE*, Vol. 53, March 1965, pp. 316–317.

10. V. L. Frantz, Indium Antimonide Thin-Film Transistor, *Proceedings of the IEEE*, Vol. 53, July 1965, p. 760.

11. J. F. Skalski, A PbTe Single-Crystal Thin-Film Transistor, *Proceedings of the IEEE*, Vol. 53, November 1965, p. 1792.

12. G. F. Boesen and J. E. Jacobs, ZnO Field-Effect Transistor, *Proceedings of the IEEE*, Vol. 56, November 1968, pp. 2094–2095.

13. D. Lile and J. C. Anderson, The Applications of Polycrystalline Layers of InSb and PbTe to a Field-Effect Transistor, *Solid State Electronics*, Vol. 12, 1969, pp. 735–741.

14. T. P. Brody and H. E. Kunig, A High-Gain InAs Thin-Film Transistor, *Applied Physics Letters*, Vol. 9, No. 7, October 1966, pp. 259–260.

15. H. E. Kunig, Analysis of an InAs Thin-Film Transistor, *Solid State Electronics*, Vol. 11, 1968, pp. 335–342.

16. H. W. Becke, R. Hall, and J. P. White, Gallium Arsenide MOS Transistors, *Solid State Electronics*, Vol. 8, 1965, pp. 812–823.

17. H. W. Becke and J. P. White, Development of a Gallium Arsenide Metal-Insulator-Semiconductor Transistor, United States Air Force Avionics Laboratory Technical Report AFAL-TR-67-117, June 1967.

18. H. W. Becke and J. P. White, A 200 MHz 300°C. Gallium Arsenide MIS Transistor, presented at the 1967 International Electron Device Meeting, Washington, D.C., October 1967.

19. A. S. Grove, *Physics and Technology of Semiconductor Devices*, John Wiley & Sons, New York, 1967, pp. 347–350.

20. N. Ditrick, M. M. Mitchell, and R. Dawson, A Low Power MOS Tetrode, presented at the 1965 International Electron Device Meeting, Washington, D.C., October 1965.

21. M. M. Mitchell, R. Dawson, and N. Ditrick, High-Frequency Characteristics of the MOS Tetrode, presented at the 1966 NEREM Conference, Boston, Massachusetts, November 1966.

22. R. H. Dawson and J. O. Preisig, MOS Dual-Gate Transistor for UHF Applications, presented at the 1969 NEREM Conference, Boston, Massachusetts, November 1969.

23. D. M. Brown, W. E. Engeler, M. Garfinkel, and P. V. Gray, Self-Registered Molybdenum-Gate MOSFET, *Journal of the Electrochemical Society*, Vol. 115, No. 8, August 1968, pp. 874–876.

24. D. M. Brown, W. R. Cady, J. W. Sprague, and P. J. Salvagni, The *p*-Channel

Refractory Metal Self-Registered MOSFET, *IEEE Transactions on Electron Devices*, Vol. ED-18, No. 10, October 1971, pp. 931–940.

25. W. J. Laughton, Molybdenum Gates Open the Door to Faster MOS Memories, *Electronics*, April 12, 1971, pp. 68–73.

26. R. W. Bower and H. G. Dill, Insulated-Gate Field-Effect Transistors Fabricated Using the Gate as a Source-Drain Mask, presented at the 1966 International Electron Device Meeting, Washington, D.C., October 1966.

27. J. C. Sarace, R. E. Kerwin, D. L. Klein, and R. Edwards, Metal-Nitride-Oxide-Silicon Field-Effect Transistors with Self-Aligned Gates, *Solid State Electronics*, Vol. 11, 1968, pp. 653–660.

28. F. Faggin, T. Klein, and L. Vadasz, Insulated-Gate Field-Effect Transistor Integrated Circuits with Silicon Gates, presented at the 1968 International Electron Device Meeting, Washington, D.C., October 1968.

29. D. L. Tolliver and C. J. Santoro, Deposited Oxide Contours in Multi-Layer Metal Circuitry, *Solid State Technology*, April 1971, pp. 32–36.

30. C. T. Naber, A Technique for Obtaining Tapered Oxide Steps in Silicon-Gate Integrated Circuits, presented at the Fall Meeting of the Electrochemical Society, Miami Beach, Florida, October 1972 (recent news paper).

31. F. Faggin and T. Klein, Silicon Gate Technology, *Solid State Electronics*, Vol. 13, 1970, pp. 1125–1144.

32. F. Morandi, The MOS Planox Process, presented at the 1969 International Electron Device Meeting, Washington, D.C., October 1969.

33. J. A. Appels, E. Kooi, M. M. Paffen, J. J. H. Schatorje, and W. H. C. G. Verkuylen, Local Oxidation of Silicon and its Application in Semiconductor Technology, *Philips Research Reports*, Vol. 25, 1970, pp. 118–132.

34. R. J. Patterson, Titanium-Aluminum Metalization for Multilayer Circuits, presented at the Fall Meeting of the Electrochemical Society, Miami Beach, Florida, October 1972.

35. R. W. Bower, H. G. Dill, K. G. Aubuchon, and S. A. Thompson, MOS Field-Effect Transistors Formed by Gate-Masked Ion-Implantation, *IEEE Transactions on Electron Devices*, Vol. ED-15, No. 10, October 1968, pp. 757–761.

36. N. E. Moyer, R. W. Bower, and H. G. Dill, High-Speed MOS Integrated Circuits Utilizing Ion-Implantation, presented at the 1969 Government Microcircuit Applications Conference, Washington, D.C., September 1969.

37. R. W. Bower, Applications of Ion-Implantation Doping to the Planar MOS Integrated Circuit Technology, presented at the 1969 NEREM Conference, Boston, Massachusetts, November 1969.

38. J. M. Shannon and J. R. A. Beale, The Use of Ion-Implantation for u.h.f. M.O.S.T.s, presented at the 1968 Solid State Device Research Conference, Boulder, Colorado, June 1968.

39. J. M. Shannon, J. Stephen, and J. H. Freeman, MOS Frequency Soars with Ion-Implanted Layers, *Electronics*, February 3, 1969, pp. 96–100.

40. K. G. Aubuchon, The Use of Ion-Implantation to Set the Threshold Voltage of MOS Transistors, presented at the International Conference on Properties and Use of M.I.S. Structures, Grenoble, France, June 17–20, 1969.

41. J. Macdougall, K. Manchester, and R. B. Palmer, Ion-Implantation Offers a Bagful of Benefits for MOS, *Electronics*, June 22, 1970, pp. 86–90.

42. P. J. Coppen, L. O. Bauer, and H. G. Dill, Ion-Implanted C-MOS Technology, presented at the 1971 Western Electronic Show and Convention, Los Angeles, California, August 24–27, 1971.

43. H. G. Dill and P. J. Coppen, Ion-Implanted MOS LSI Circuits, presented at the 1971 IEEE Convention, New York, March 1971.

44. R. Crawford, Implanted Depletion Loads Boost MOS Array Performance, *Electronics*, April 24, 1972, pp. 85–90.

45. C. C. Mai, M. Hswe, and R. B. Palmer, Ion-Implantation Combined with Silicon-Gate Technology, *IEEE Transactions on Electron Devices*, Vol. ED-19, No. 11, November 1972, pp. 1219–1221.

46. T. Masuhara, M. Nagata, and N. Hashimoto, A High-Performance n-Channel MOS-LSI Using Depletion-Type Load Elements, presented at the 1971 International Solid State Circuits Conference, Philadelphia, Pennsylvania, February 1971.

47. F. M. Wanlass and C. T. Sah, Nanowatt Logic Using Field-Effect Metal-Oxide-Semiconductor Triodes, presented at the 1963 International Solid State Circuits Conference, Philadelphia, Pennsylvania, February 1963.

48. T. Klein, Technology and Performance of Integrated Complementary MOS Circuits, *IEEE Journal of Solid State Circuits*, Vol. SC-4, No. 3, June 1969, pp. 122–130.

49. K. K. Yagura, G. M. Catlin, and J. D. Hutchenson, Monolithic MOS Complementary Pairs, presented at the 1966 International Electron Device Meeting, Washington, D.C., October 1966.

50. T. Athanas, Development of Low Threshold Voltage COS/MOS Integrated Circuits, presented at the 1971 IEEE Convention, New York, March 1971.

51. M. H. White and J. R. Cricchi, Complementary MOS Transistors, *Solid State Electronics*, Vol. 9, 1966, pp. 991–1008.

52. R. R. Burgess and R. G. Daniels, Silicon-Gate CMOS Integrated Circuits, presented at the 1970 International Electron Device Meeting, Washington, D.C., October 1970.

53. P. J. Coppen, K. G. Aubuchon, L. O. Bauer, and N. E. Moyer, A Complementary MOS 1.2 Volt Watch Circuit Using Ion-Implantation, *Solid State Electronics*, Vol. 15, 1972, pp. 165–175.

54. H. M. Manasevit and W. I. Simpson, Single Crystal Silicon on a Sapphire Substrate, *Journal of Applied Physics*, Vol. 35, April 1964, pp. 1349–1351.

55. C. W. Mueller and P. H. Robinson, Grown-Film Silicon Transistors on Sapphire, *Proceedings of the IEEE*, Vol. 52, No. 12, December 1964, pp. 1487–1490.

56. E. C. Ross and C. W. Mueller, Extremely-Low Capacitance Silicon Film MOS Transistors, *IEEE Transactions on Electron Devices*, Vol. ED-13, March 1966, pp. 379–381.

57. S. R. Hofstein, An Analysis of Deep-Depletion Thin-Film MOS Transistors, *IEEE Transactions on Electron Devices*, Vol. ED-13, No. 12, December 1966, pp. 846–855.

58. F. P. Heiman, Thin-Film Silicon-on-Sapphire Deep-Depletion MOS Transistors, *IEEE Transactions on Electron Devices*, Vol. ED-13, No. 12, December 1966, pp. 855–862.

59. J. F. Allison, F. P. Heiman, and J. R. Burns, Silicon-on-Sapphire Complementary MOS Memory Cells, *IEEE Journal of Solid State Circuits*, Vol. SC-2, No. 4, December 1967, pp. 208–212.

60. J. F. Allison, J. R. Burns, and F. P. Heiman, Silicon-On-Sapphire Complementary

MOS Memory Systems, presented at the 1967 International Solid State Circuits Conference, Philadelphia, Pennsylvania, February 1967.

61. J. R. Burns and J. H. Scott, Silicon-on-Sapphire Complementary MOS Circuits for High-Speed Associative Memory, presented at the 1969 Fall Joint Computer Conference, Las Vegas, Nevada, November 1969.

62. J. E. Meyer, J. R. Burns, and J. H. Scott, High-Speed Silicon-on-Sapphire 50 Stage Shift Register, presented at the 1970 International Solid State Circuits Conference, Philadelphia, Pennsylvania, February 1970.

63. E. J. Boleky, The Performance of Complementary MOS Transistors on Insulating Substrates, *RCA Review*, Vol. 31, No. 2, June 1970, pp. 372–395.

64. G. W. Cullen, G. E. Gottlieb, and J. H. Scott, The Epitaxial Deposition of Silicon on Insulating Substrates for MOS Circuitry, presented at the Technical Conference on Preparation and Properties of Electronic and Magnetic Materials for Computers, The Metallurgical Society, New York, August 31–September 2, 1970.

65. A. K. Rapp and E. C. Ross, Silicon-on-Sapphire Substrates Overcome MOS Limitations, *Electronics*, September 25, 1972, pp. 113–116.

66. G. W. Cullen, G. E. Gottlieb, and C. C. Wang, The Epitaxial Growth of Silicon on Sapphire and Spinel Substrates: Suppression of Changes in the Film Properties During Device Processing, *RCA Review*, Vol. 31, No. 2, June 1970, pp. 355–371.

67. H. M. Manasevit and D. H. Forbes, Single-Crystal Silicon on Spinel, *Journal of Applied Physics*, Vol. 37, February 1966, p. 734.

68. P. H. Robinson and D. J. Dumin, The Deposition of Silicon on Single-Crystal Spinel Substrates, *Journal of the Electrochemical Society*, Vol. 115, January 1968, p. 75.

69. K. H. Zaininger and C. C. Wang, MOS and Vertical Junction Device Characteristics of Epitaxial Silicon on Low Aluminum-Rich Spinel, *Solid State Electronics*, Vol. 13, No. 7, July 1970, pp. 943–950.

70. M. T. Duffy, C. C. Wang, and G. W. Cullen, Preparation and Properties of MIS Structures in Silicon-on-Spinel Using Thin Films of Al_2O_3 and Si_3N_4, presented at the Technical Conference on Preparation and Properties of Electronic and Magnetic Materials for Computers, The Metallurgical Society, New York, August 31–September 2, 1970.

71. E. J. Ryder, Mobility of Holes and Electrons in High Electric Fields, *Physical Review*, Vol. 81, June 1953, pp. 766–769.

72. C. B. Norris and J. F. Gibbons, Measurement of High-Field Carrier Drift Velocities in Silicon by a Time-of-Flight Technique, *IEEE Transactions on Electron Devices*, Vol. ED-14, No. 1, January 1967, pp. 38–43.

73. V. Rodriquez, H. Ruegg, and M. A. Nicolet, Measurement of the Drift Velocity of Holes in Silicon at High Field Strengths, *IEEE Transactions on Electron Devices*, Vol. ED-14, No. 1, January 1967, pp. 44–46.

74. S. R. Hofstein, Field-Effect Transistor Theory, *Field-Effect Transistors, Physics, Technology and Applications*, edited by J. T. Wallmark and H. Johnson, Prentice-Hall Inc., Englewood Cliffs, New Jersey, 1967, pp. 125–130.

75. F. F. Fang and B. L. Crowder, Ion-Implanted Microwave MOSFETs, presented at the 1970 International Electron Device Meeting, Washington, D.C., October 1970.

76. R. E. Harris, Double-Diffused MOS Transistors, presented at the 1967 International Electron Devices Meeting, Washington, D.C., October 1967.

77. Y. Tarui, Y. Hayashi, and T. Sekigawa, DSA Enhancement-Depletion MOS I.C.,

presented at the 1970 International Electron Devices Meeting, Washington, D.C., October 1970.

78. H. C. Lin and C. J. Varker, Normally-On Load Device for IGFET Switching Circuits, presented at the 1969 Northeast Electronics Research and Engineering Meeting (NEREM), Boston, Massachusetts, November 1969.

79. T. P. Cauge and J. Kocsis, A Double-Diffused MOS Transistor with Microwave Gain and Sub-Nanosecond Switching Speeds, presented at the 1970 International Electron Devices Meeting, Washington, D.C., October 1970 (late news paper).

BIBLIOGRAPHY

Appels, J. A., and M. M. Paffen, Local Oxidation of Silicon; New Technological Aspects, *Philips Research Reports*, Vol. 26, 1971, pp. 157–165.

Boleky, E. J., Subnanosecond Switching Delays Using CMOS/SOS Silicon-Gate Technology, presented at the 1971 International Solid State Circuits Conference, Philadelphia, Pennsylvania, February 1971.

Brown, D. M., The Self-Registered MOSFET—A Brief Review, *Solid State Technology*, April 1972, pp. 33–45.

Burgess, R. R., and R. G. Daniels, C/MOS Unites with Silicon Gate to Yield Micropower Technology, *Electronics*, August 30, 1971, pp. 38–43.

Burns, J. R., High-Frequency Characteristics of the Insulated-Gate Field-Effect Transistor, *RCA Review*, Vol. 28, No. 3, September 1967, pp. 385–418.

Cauge, T. P., J. Kocsis, H. J. Sigg, and G. D. Vendelin, Double-Diffused MOS Transistor Achieves Microwave Gain, *Electronics*, February 15, 1971, pp. 99–104.

Critchlow, D. L., The n-Channel IGFET for Logic and Memory, presented at the 1968 Northeast Electronics Research and Engineering Meeting (NEREM), Boston, Massachusetts, November 1968.

Dill, H. G., R. M. Finnila, A. M. Leupp, and T. N. Toombs, The Impact of Ion Implantation on Silicon Device and Circuit Technology, *Solid State Technology*, December 1972, pp. 27–35.

Edwards, J. R., and G. Marr, Depletion-Mode IGFET Made by Deep Ion-Implantation, *IEEE Transactions on Electron Devices*, Vol. ED-20, No. 3, March 1973, pp. 283–289.

Eversteyn, F. C., and H. L. Peek, Preparation and Stability of Enhancement n-channel MOS Transistors with High Electron Mobility, *Philips Research Reports*, Vol. 24, 1969, pp. 15–33.

Hswe, M., R. B. Palmer, M. L. Shopbell, and C. C. Mai, Characteristics of p-Channel MOS Field-Effect Transistors with Ion-Implanted Channels, *Solid State Electronics*, Vol. 15, 1972, pp. 1237–1243.

Kalinowski, J. J., A Compact MOST Model for Design Analysis, *Proceedings of the IEEE*, Vol. 60, No. 8, August 1972, pp. 1000–1001.

Kooi, E., J. G. van Lierop, W. H. C. G. Verkuijlen, and R. de Werdt, LOCOS Devices, *Philips Research Reports*, Vol. 26, 1971, pp. 166–180.

Lambrechtse, C. W., R. H. W. Salters, and L. Boonstra, A 4K 1-MOS/Bit RAM with Internal Timing and Low Dissipation, presented at the 1973 International Solid State Circuits Conference, Philadelphia, Pa., February 1973.

Lin, H. C., and W. N. Jones, Computer Analysis of the Double-Diffused MOS Transistor for Integrated Circuits, *IEEE Transactions on Electron Devices*, Vol. ED-20, No. 3, March 1973, pp. 275–283.

Lin, H. C., J. L. Halsor, and P. J. Hayes, Shielded Silicon Gate Complementary MOS Integrated Circuit, *IEEE Transactions on Electron Devices*, Vol. ED-19, No 11, November 1972, pp. 1199–1207.

MacPherson, M. R., Threshold Shift Calculations for Ion-Implanted MOS Devices, *Solid State Electronics*, Vol. 15, 1972, pp. 1319–1326.

McLintock, G. A., R. E. Thomas, and R. S. C. Cobbold, Modelling of Double-Diffused MOSTs with Self-Aligned Gates, presented at the 1972 International Electron Devices Meeting, Washington, D.C., December 1972.

McLintock, G. A., R. E. Thomas, R. S. C. Cobbold, and J. G. Hogeboom, Forward and Reverse Characteristics of Self-Aligned Double-Diffused M.O.S. Transistors, *Electronics Letters*, Vol. 8, No. 18, September 7, 1972, pp. 463–465.

Mansour, I. R. M., E. A. Talkhan, and A. I. Barboor, Investigations on the Effect of Drift-Field-Dependent Mobility on MOST Characteristics, *IEEE Transactions on Electron Devices*, Vol. ED-19, No. 8, August 1972, pp. 899–916.

Morandi, F., Planox Process Smoothes Path to Greater MOS Density, *Electronics*, December 20, 1971, pp. 44–48.

Palmer, R. B., Ion-Implantation Technology, presented at the Second Annual IEEE Metropolitan New York Electron Device Group Symposium on MOS Devices and Integrated Circuits, New York, April 20, 1972.

Reddi, V. G. K., and A. Y. C. Yu, Ion-Implantation for Silicon Device Fabrication, *Solid State Technology*, October 1972, pp. 35–41.

Richman, P., Suppression of Parasitic Thick-Field Conduction Mechanisms in Silicon-Gate M.O.S. Integrated Circuits, *Electronics Letters*, Vol. 7, No. 1, January 14, 1971, pp. 12–13.

Richman, P., *n*-Channel Coplamos Technology, presented at the Second Annual IEEE Metropolitan New York Electron Device Group Symposium on MOS Devices and Integrated Circuits, New York, April 20, 1972.

Sigg, H. J., G. D. Vendelin, T. P. Cauge, and J. Kocsis, D-MOS Transistor for Microwave Applications, *IEEE Transactions on Electron Devices*, Vol. ED-19, No. 1, January 1972, pp. 45–53.

PROBLEMS

7.1 For a bipolar *n-p-n* transistor, if the collector junction leakage current is small with the emitter open-circuited, the collector current as a function of the emitter current will be approximately

$$I_C \cong \alpha I_E,$$

where

$$I_E \cong \frac{q D_{nB} n_{pB} A_E}{X_B} \left[\exp\left(\frac{q V_{BE}}{kT}\right) - 1 \right]$$

for moderate injection levels. The D_{nB} is the diffusion constant for electrons in the base region, n_{pB} is the equilibrium concentration of electrons per unit volume in the base, and X_B is the width of the base region. Assuming that $(qV_{BE}/kT) \gg 1$, show that the transconductance of the device will be approximately equal to

$$g_m \cong \frac{qI_E}{kT} \, .$$

Using material values and dimensions typical of bipolar and MOSFET integrated circuit structures, compare the transconductance per unit area of an *n-p-n* bipolar transistor with that of an *n*-channel MOSFET as a function of collector and drain currents, respectively, at $T = 25°C$.

7.2 Derive an expression for the incremental circuit gain as a function of frequency when the simple amplifier shown in Figure 7.3 is used to drive a grounded capacitive load, C_L. If C_{GD}, C_{GS}, and C_{DB} are extremely small for the transistor used in the amplifier, express the incremental circuit gain as a function of frequency in terms of G_0, the low-frequency incremental circuit gain.

7.3 Show that, for drain voltages below pinch-off, the maximum operating frequency of a perfectly self-aligned IGFET structure is approximately equal to

$$f_{\max} \cong \frac{\mu \, |V_D|}{2\pi L^2} = \frac{g'_m}{2\pi C_{ch}} \, ,$$

where g'_m is the transconductance below pinch-off as a function of gate voltage.

7.4 For the dual-gate MOS tetrode structure illustrated in Figure 7.8, discuss the effect on device gain and high-frequency performance when, with the source grounded, the input signal is applied at gate 2 and the fixed d.c. bias is applied at gate 1.

7.5 Discuss the effect of reverse substrate bias upon the electrical characteristics of high-impedance ion-implanted resistors. Compare and contrast their behavior under reverse substrate bias conditions with conventional diffused resistors and with evaporated thin film resistors. Discuss the role played by the profile of the implanted impurity concentration.

7.6 Calculate the respective threshold voltages of the *n*- and *p*-channel (deep-depletion) MOSFETs fabricated on a *p*-type silicon-on-sapphire film with $N_A = 1 \times 10^{15}/cm^3$ and a thickness equal to $0.7 \, \mu$. The thickness of the silicon dioxide gate insulator is equal to 1200 Å, aluminum is used for the gate electrodes, and $Q_{SS}/q = 5 \times 10^{10}/cm^2$.

7.7 Calculate the respective threshold voltages of the *n*- and *p*-channel MOSFETs fabricated through the use of a two-stage epitaxial deposition process on *p*- and *n*-type silicon-on-sapphire films with $N_A = 1 \times 10^{15}/cm^3$ and $N_D = 1 \times 10^{15}/cm^3$, respectively. Both films have thicknesses equal to $0.7 \, \mu$ and the thickness of the silicon dioxide gate insulator is equal to 1200 Å. Aluminum is used for the gate electrodes and $Q_{SS}/q = 5 \times 10^{10}/cm^2$.

7.8 Discuss how one would go about predicting the threshold voltage of a diffusion-self-aligned MOSFET. Consider the profiles of the p^- and n^+ diffusions in the lateral direction beyond the source window (for the case of an n-channel device). Where will the effective surface acceptor doping concentration be the highest? Where will pinch-off occur first?

List of Symbols

C_{ch}	Capacitance from the gate electrode to the active channel
C_{DB}	Drain-to-substrate capacitance
C_{dct}	Total effective coupling capacitance between drain and channel
C_{GD}	Gate-to-drain negative feedback capacitance
C_{GS}	Gate-to-source capacitance
C_{in}	Gate input capacitance
C_{ox}	Dielectric capacitance per unit area across the insulating gate oxide
C_s	Capacitance per unit area associated with the surface space charge region
C_{SB}	Source-to-substrate capacitance
C_{SD}	Capacitance per unit area associated with the surface depletion region
$C_{SD_{\min}}$	Minimum value of the capacitance per unit area associated with the surface depletion region
C_{ST}	Capacitance per unit area associated with charged fast states
C_{tox}	Total oxide capacitance
D_{DB}	Drain-to-substrate diode symbol
D_{SB}	Source-to-substrate diode symbol
E_G	Band-gap energy
$\mathcal{E}$	Electric field intensity
$\mathcal{E}_{ox}$	Electric field within the gate oxide layer
$\mathcal{E}_s$	Electric field at the silicon surface
f	Frequency
$f_{\max}$	Maximum operating frequency
G	Amplifier gain
$g_{D_{sat}}$	Conductance in the region of saturated drain current
g_{DSO}	Drain-to-source conductance with zero drain voltage
g_m	Transconductance
I_D	Drain current
$I_{D_{sat}}$	Saturated drain current
I_{DW}	Drain current per unit channel width
I_{SCL}	Space-charge-limited current
J	Current density

251

k	Boltzmann's constant
L	Drain-to-source separation
L_D	Intrinsic Debye length
m	Segregation coefficient
m_e	Effective mass of an electron
n	Electron concentration per unit volume
n_i	Intrinsic concentration per unit volume
n_s	Electron concentration per unit volume at the surface of the semiconductor
N_A	Acceptor doping concentration per unit volume
N_A^-	Ionized acceptor concentration per unit volume
N_D	Donor doping concentration per unit volume
N_D^+	Ionized donor concentration per unit volume
N_{ST}	Number of charged fast states per unit volume
p	Hole concentration per unit volume
p_s	Hole concentration per unit volume at the surface of the semiconductor
q	Electronic charge
Q_{av}	Average charge density per unit area in the conducting portion of the channel
Q_G	Charge density per unit area present on the gate electrode
Q_{inv}	Charge density per unit area in the surface inversion layer
Q_n	Electron density per unit area in an n-type surface inversion layer
Q_p	Hole density per unit area in a p-type surface inversion layer
Q_{SD}	Charge density per unit area associated with the surface depletion region
$Q_{SD_{max}}$	Maximum value of the charge density per unit area associated with the surface depletion region
$Q'_{SD_{max}}$	Maximum value of the charge density per unit area associated with the surface depletion region as a function of the applied substrate-to-source voltage
Q_{SS}	Fixed positive charge density per unit area at the silicon-silicon dioxide interface
Q_{total}	Total charge density per unit area
R	Resistance
R_D	Dynamic drain-to-source resistance
R_{DS}	Drain-to-source resistance

$r_{D_{sat}}$	Drain saturation resistance
R_L	Load resistance
R_p	Mean projected range associated with Gaussian implantation
ΔR_p	Mean standard deviation associated with Gaussian implantation
t	Transit time
T	Temperature, in degrees Kelvin
T_{ox}	Thickness of the insulating gate oxide layer
v_d	Drift velocity
$v_{d_{av}}$	Average drift velocity
V_D	Drain voltage
$V_{D_{sat}}$	Drain voltage at which current saturation occurs
V_{FB}	Flatband voltage
V_G	Gate voltage
v_n	Electron velocity
V_0	Contact potential
V_{OX}	Voltage across the oxide layer
V_{pt}	Drain-to-source punch-through voltage
V_S	Applied substrate-to-source voltage
V_T	Threshold voltage
ΔV_T	Change in threshold voltage as a function of the applied substrate-to-source potential
v_1	Saturated carrier drift velocity
W	Channel width
x_d	Width of the surface depletion region
$x_{d_{max}}$	Maximum width of the surface depletion region
$x_{silicon}$	Silicon film thickness
β	MOSFET gain factor
ϵ_{ox}	Dielectric constant of the oxide gate insulating layer
ϵ_s	Dielectric constant of silicon
λ	Mean free path between collisions
μ	Mobility
μ_n	Electron mobility
μ_p	Hole mobility
ρ	Space charge density per unit volume
σ	Total interface charge density per unit area
σ_C	Conductivity

τ	Relaxation time
τ_D	Dielectric relaxation time
τ_{inv}	Surface inversion layer response time
ϕ	Electrostatic potential
ϕ_F	Fermi potential
ϕ_{MOX}	Metal-SiO$_2$ barrier energy
$\phi_{MS'}$	Metal-semiconductor work function difference
ϕ_s	Surface potential
ϕ_{s0}	Surface potential with zero gate voltage
ϕ_{SOX}	Silicon-SiO$_2$ barrier energy
ω_{max}	Maximum operating frequency

Index

255